CHARACTERISATION OF F
MATERIA

Series Editor: M McLEAN

NUMERICAL
TECHNIQUES

*Proceedings of the seventh
seminar in a series
sponsored and organised by the
Materials Science, Materials Engineering and
Continuing Education Committees of
The Institute of Metals;
held in London on 6 December 1989*
EDITED BY P SPILLING

**THE INSTITUTE OF METALS
1989**

Book 449
Published in 1989 by
The Institute of Metals
1 Carlton House Terrace, London SW1Y 5DB
and
The Institute of Metals
Morth American Publications Center
Old Post Road, Brookfield, VT 05036 U S A

I S B N 0 901462 68 3

Compiled from typematter and illustrations provided by the contributors

Made and printed in Great Britain
by M & A Thomson Litho Ltd, East Kilbride, Scotland

CONTENTS

PREFACE

This is the seventh and last volume in the series 'Characterisation of High Temperature materials'. It attempts to review some of the many areas in which numerical methods can be applied as basic tools for the solution of metallurgical problems and to provide a grounding in the principles involved. The powerful numerical tools that are being evolved will fundamentally change the way that a metallurgist or material scientist approaches a problem. The accuracy and usefulness of any numerical simulation is dependant on the generation or availability of quality, quantitative data for the material properties of interest. A greater emphasis, therefore, will be placed on the systematic generation of databases upon which these numerical 'tools' can operate.

In the previous volumes of this series techniques for the characterisation of materials have been reviewed. It is the purpose of this book to show how that data can be analysed, organised and subsequently applied to complex problems. It is, therefore, appropriate that the first paper introduces 'Image Analysis' since the development of this technique renders an area, previously regarded as essentially subjective, accessible to analysis. This ability is critical to future effective modelling in areas of microstructural development. The subsequent papers cover the formation of databases, data analysis and an introduction to numerical analysis techniques. Finally the two concluding papers bring together the methods outlined in the preceding sections and show practical applications of these techniques to material behaviour and process modelling.

Paul Spilling
Manufacturing Technology
Rolls-Royce
Derby

All papers are printed
as received from the
authors and arranged
with Tables and Figures
following text

SESSION 1
Chairman: P Spilling
(Rolls Royce Plc, Derby)

1:Quantification of microstructures and surface topography

E R WALLACH

University of Cambridge

1. INTRODUCTION

Within the last ten years, the information technology revolution has come about as a result of, and also has been fuelled by, the increased accessibility of computing power. Most scientific instruments, industrial equipment and domestic appliances now rely on computers to operate, and new computers are cheaper, readily available and user-friendly. Thus in scientific and engineering applications, quantification is now used routinely whereas, before it might have been time-consuming and expensive. There are many scientific techniques that are benefitting from these developments including image analysis, the subject of this chapter and typical of one of the many quantitative techniques now used in materials' science. While traditionally used in materials' science to classify microstructures and so to help establish microstructure-property relationships, image analysis also has widespread usage in such diverse fields as data collection/processing for driverless vehicles, enhancement of images for the media, and the recognition of shapes and sizes in automated quality control and security systems.

   An objective of image analysis in materials' science is to be able to measure parameters from one or more fields of view obtained from sources such as photographs and direct microscope images. This can readily be achieved either by using a video camera to capture an image or by processing the imaging signal directly from a microscope, and in either case then feeding the resulting signal via an interface to an image analyser. The image is thus digitised into small discrete areas (picture elements or pixels), each of which has a measured brightness (intensity or, in mono-chromatic systems, shades of grey). To a first approximation, the greater the number of pixels and available intensity levels, the higher the resolution of the digitised image and the better the potential discrimation of features within the image. Following digitisation, the image can be enhanced, if required, and then analysed. Note that a distinction should be made at the onset between image analysis (the measurement of features within a field of view) and image enhancement (manipulation of the original input digitised image in order to improve the appearance of the image and/or to facilitate the detection of features). The data remain unaffected in the former whereas the data (values of intensity for pixels and/or the number of pixels defining a feature) may be altered considerably in the latter, hence possibly affecting what subsequently is measured.

Having identified and measured features of interest, the final stages are statistical interpretation and stereology (mathematical manipulations used to transform the two-dimensional data acquired back into three-dimensional representations characteristic of the original microstructure).

This chapter considers the various stages of image analysis. Further details can be found in the references at the end of the chapter.

2. INSTRUMENTATION

A typical image analyser comprises the following:

- image sensor (video camera or signal from electron microscope);
- memory or frame store to store the image data;
- processor or computer to enhance and/or analyse the image;
- monitor to display operations and images;
- input and output devices (keyboard and/or mouse, printer);
- optional data/results storage device (floppy or hard disc drive).

Aspects of some of these hardware components are outlined in this section; the remaining items are standard computing accessories.

2.1 Video cameras

Many image analysers are stand-alone systems utilising a video camera to capture the image to be processed. Until recently, electron tube cameras were the most common but in recent years developments in solid-state cameras (of which the most common are the CCD, charge coupled device) have been such that they now are becoming competitive.

Electron tube cameras monitor the local resistance of a photosensitive, conducting target or face plate onto which the light from the original image falls. This is accomplished using an electron beam to raster the back of the target and then measuring the magnitude of the local current which flows through the target to a thin transparent conductive layer (e.g. tin oxide) on the front of the target. The amount of light reaching different parts of the photosensitive face plate changes the resistance locally and so the transmitted current can be correlated with regions of differing brightness on the original image.

3

The signal is then amplified and digitised to provide the input to the image analyser and an electronic representation of the original image.

There are a number of types of tube camera which differ mainly in the materials used for their photoconductive layers and hence in their relative spectral sensitivities and response to incident light. Two other properties of relevance to the performance of cameras are:

- lag - the time required for the signal to decay after removal of the illumination, ideally as short as possible especially if a number of images (or fields) per second are to be collected in an automated system;

- bloom - an undesired increase in detected area compared with the real area of the viewed feature as a consequence of the brightness of the object).

The most commonly used tube has been the vidicon (typically an antimony trisulphide photoconductive layer); properties of this and other tube cameras are summarised below.

Tube type	Spectral response	Tube response to incoming light	Degree of lag
vidicon	as eye	output decreases as light increases	large
newvicon	low blue; peaks in red	high linear output signal	large
plumbicon	low red	blooms in bright light	low
chalnicon	linear	very high linear output signal	large
saticon	low red	low output signal	low

[Note: further details are given in Wezel [1].]

There are two major disadvantages of tube cameras, namely their relatively poor geometrical accuracy in recording an image and their lack of uniformity of response to incoming light. The poor geometric accuracy comes, in part, from variablities in the scan circuits which result in geometric distortions such that the same object imaged by different parts of a sensor would have different sizes, e.g. "pin-cushioning" whereby the centre of an image appears less magnified than the edges. In practice, the effect can be minimised in image analysis applications as follows. Since distortions may also occur in the imaging lenses, it can be useful to calibrate a given camera together with a particular lens in a microscope by imaging a rectangular grid pattern. The resulting image (incorporating the distortions from both lens and camera tube) can be stored in some image analysers and then used to correct any subsequent images collected by the same camera and lens (clearly, a change in magnification, hence a different objective lens with its own particular inaccuracies, would require a different correction pattern).

The second disadvantage of tube cameras is a variation in sensitivity to the incoming light; higher outputs are observed in the centre of the face plate than at the edges.

Solid-state cameras do not have tubes and electron beams. Instead, these cameras utilise a matrix of sensors or cells (one or more of which will correspond to each pixel) on an electronic chip, and whose individual responses are related to the intensity of the incoming light at each location on the chip. Thus the signal is read out sequentially from the sensors on the chip. A variety of electronic devices have been used to form these sensors; MOS capacitors, photodiodes or photo-conductors.

While these cameras used to have very short life-times (in some cases, less than one year), their development now is such that they are replacing vidicon tube cameras. Their advantages are their robust nature, their good geometrical accuracy and their uniform response to the incoming light intensity (in fact, while individual sensors on the same chip may respond slightly differently to each other, a calibration under standard conditions can be stored and used to correct subsequent images). However, due to their mode of operation, they have a poor spectral sensitivity to blue light (which penetrates less well), yet a good response into the infra-red (requiring the use of infra-red filters in some applications). The CCD camera is being used increasingly in image analysers and is the camera which will be used predominantly in the future.

Video is the analogue signal that is produced by the camera, and contains electronic data that will be decoded in the image analyser to form the image. Unfortunately, the commonly accepted Comité Consultatif International de Radiodiffusion [CCIR] "standard" varies from country to country and thus care must be taken to ensure that a camera and image analyser are suitably matched. In the U.K., the standard is CCIR I (PAL) for colour television and requires that the image has 625 lines (transmitted at 50 Hz in two groups or "fields" of 312.5 lines, i.e. each line scan taking 20 ms) in an "interlaced" manner. A typical signal has a maximum output of 1 volt with the image data ranging from 0.3 V (black) to 1.0 V (white) The part of the signal from 0 to 0.3 V contains control or synchronisation pulses such as those which indicate the beginning of each of the 625 lines, and blanking pulses which turn off the screen while the scan moves back to the beginning of the next line.

2.2 Processor

Processors may be dedicated hardware or micro-computers with appropriate programs. Either can be used to control the collection of data (including stage movements and auto-focussing in fully automated systems), to process data (contrast enhancement, feature identification and editting), to make accurate measurements, and to statistically analyse and interpret results. Software is written and supplied by manufacturers although it is possible in many systems for users to write and use their own subroutines; this can be essential if the types of image to be analysed vary and are not fixed (as in production or quality control usages).

Analogue to digital (A/D) conversion is used to convert the analogue video signal into a digital form so that the data can be handled by the processor. The intensity of the original signal is represented by a number usually between 0 (black) and 255 (white), i.e. there are 256 grey shades (equivalent to 2^8 or 8 bit resolution); note that the eye can just about resolve this number of intensity levels. Spatial resolution is governed by the number of pixels that can be stored; images are typically 256 or 512 pixels square. The data so obtained are stored in a memory or *frame store*.

The desire for more grey levels and better spatial resolution must be tempered by the need to maintain reasonable speed when handling the increased amount of data and by the computer or procesor memory needed just to store it. An image comprising 512 by 512 pixels and with 256 grey scales requires memory storage of 256 kilo-bytes, a sizable proportion of a personel computer memory.

Look-up tables. Having obtained a digital signal, image enhancement and image analysis can be undertaken (see section 4). These are achieved by defining various operations to be carried out on the intensity value associated with each pixel. Typically, the intensity or grey scale for an image may have a range of up

7

to 256 possible values (0 corresponding to black and 255 to white). There also may be 512^2 (i.e. over 260,000) pixels defining the spatial representation of an image. Therefore, it is generally faster to calculate and store 256 new pixel values for a particular operation selected and then *"look-up"* the new value for each pixel rather than *calculate* a revised intensity value for each of the 262144 pixels. To optimise speed, this operation is often carried out using dedicated hardware rather than within the memory of a computer, generally a slower process. The speed of these calculations is important if operations are to be viewed in real-time.

2.3 Display and output of results

TV display. Image analysis is a subjective process, for instance one or more of a number of quite specific features in an image may need to be identified. Hence interaction with the operator is essential and it is necessary to display both the original digitised image and the consequences of any processing instructions undertaken (e.g. filters applied); this is done by a D/A conversion, i.e. a reverse of the above A/D operation, to form a video signal which can be displayed on a TV monitor. It is usual for this to be a colour monitor so that colours can be used to indicate, for example, detected features and so facilitate feature identification.

Results of image analysis are often sets of numbers corresponding to the types of features present in an image. These may be displayed on the TV display and printed out as tables or in graphical form, e.g. histograms to show population distributions or scatter graphs to show the extent of correlations between two variables in sets of measurements. Results also may be stored, together with the original image, on floppy or hard disc and can be tranferred directly to another computer for further manipulation, statistical analysis, incorporation into data bases, etc.

3. DATA COLLECTION

3.1 Sample

Before attempting any analysis, one has to be sure one knows what is to be measured and why; the nature of the investigation must be understood before a sample is prepared. For instance, if attempting to measure the grain size of an extruded rod, the orientation of any metallographic cross-section with respect to the extrusion direction must be known if the subsequent results are to have any meaning; small equiaxed grains would be seen if viewed end-on, whereas long elongated grains would be observed in a longitudinal section. As a general rule, sections from three mutually perpendicular orientations are advisable when first investigating a problem.

Having decided on the type of sample and a representative method of sectioning, the subsequent preparation will be of relevance. The dangers of introducing a bias into the results as a consequence of etching have been illustrated in an earlier article [2] in this series; for instance, the relative sizes of different phases and/or inclusions will change if the etch does not attack all constituents of the structure equally. Also, since etches are in general selective, the selection of an unsuitable etch may result in a relevant feature not being observed at all unless the problem has been properly defined and an appropriate proceedure adopted. In a similar way, the measurement of voids can be affected by sample preparation; the polishing of soft materials may result in smearing over voids so that they look smaller than their original sizes, and etches can selectively remove material at the edges of voids so that they appear larger.

9

3.2 Image generation

(a) Technique used

Experimental techniques with which microstructural data can be collected have been covered in an earlier book [3] in this series. Light microscopy, scanning and transmission electron microscopies are the most widely used although there are many variants within these headings. Again, the ability to subsequently measure features accurately in an image analyser will be governed by the original quality of the image. As an example, consider deformation twins in zinc; Fig. 1 shows a reflected light micrograph under normal illumination and the same region viewed in polarised light using crossed polars. While a faint image of the deformation twins can just be seen under normal illumination (Fig. 1a), the clarity of the images viewed in polarised light ensures that any image enhancement required later will be minimal. On the other hand, even this simple example introduces its own problem; rotation of the sample with respect to the crossed polars alters the image seen and hence the information processed (compare Figs. 1b and 1c) and so care would have to be taken to ensure that sufficient representative images were collected.

(b) Magnification

As described above, the nature of the problem to be analysed influences the selection and sectioning of the sample. So too does the choice of magnification. For instance, if you look out of a window, features such as chimneys are readily seen on nearby houses whereas they are not distinguishable on houses in the distance; accordingly, if the outside perimeter of a house was measured, the magnification would affect the result since different features are included in the two cases. In addition, the percentage accuracy of the result can be influenced by the magnification; at high magnification (in the example used, the nearby

10

house), the actual measurement will of a large number of pixels with a relatively higher percentage accuracy than for low magnification (distant house and correspondingly smaller number). In general, individual features should comprise at least 30 pixels each. Less than this number can mean that the precise position of a feature within the image can lead to a large percentage variation in the number of pixels representing that feature; subsequent measurements would be liable to high scatter and a loss of precision.

The effect of magnification on images of a gear wheel is illustrated in Fig. 2 and this is discussed further in section 5.2.

(c) Image Distortion
Most imaging systems have inherent distortions or aberrations associated with them which can introduce systematic errors into any subsequent analyses of images. In critical applications, it is possible to use control specimens of known symmetry to investigate the extent of any distortion (the use of a rectangular grid pattern for calibrating a microscope lens and video cameras is one example - discussed in section 2.1). However, where possible, it is better to minimise any distortion by optimising the instrumental conditions or set-up.

One also has to be aware of the limitations of the various image enhancement features that may be available on a microscope. An example of this is shown in Fig. 3; the schematic diagram shows how a sphere on a rectangular grid might appear in a scanning electron micrograph. If viewed directly from above, the only distortions would be those from the microscope and both sphere and grid would look reasonable. However, if the sample is tilted at 45º, then the grid would appear distorted, see Fig. 3b. For those familiar with image manipulation in scanning electron micrographs, one reaction might be to use an electronic tilt correction available on some instruments. The result would be as shown in

11

Fig.3c; the grid now appears square again but, not surprisingly, the sphere is now distorted! The only solution would be to revert to viewing at normal incidence.. While the distortions are obvious in this example, in many cases such image distortions will not be obvious and incorrect measurements may be obtained.

Finally, photographic prints are often presented for analysis. It is instructive to photograph a symmetrical rectangular grid using a microscope with minimal aberrations. The final print may well show distortions due to the uneven contractions of the photographic paper during printing; a related example is the image distortion introduced when photocopying. Hence care must be exercised both when preparing images and also when preparing calibration grids; the use of acetate film which does not exhibit uneven contractions can be advantageous, if only for the preparation of calibration grids used to correct lens and camera aberrations.

(d) Summary
The importance of the issues covered in this section cannot be stressed enough. Despite long pre-dating computers, the well-known expression "you cannot make a silk purse out of a sow's ear" is particularly applicable to image analysis; only if the image is appropriate to the problem and is of reasonable quality can image analysis provide meaningful data. Moreover, a great deal of time can be spent trying to recover information from a poor image using various enhancement methods when, instead, a more appropriate means of collecting the image in the first instance would immediately minimise difficulties and could improve accuracy.

4. IMAGE ENHANCEMENT AND ANALYSIS

Having optimised the setup for data collection
from representative samples, several steps may
be required in order to carry out image
analysis and so to obtain the required data,
namely:

(a) corrections to the image (i.e. minimisation
    of artefacts from illumination, lens and/or
    camera) and image calibration ;
(b) enhancement to facilitate detection of
    features within an image;
(c) measurement and interpretation of data.

These steps, the last two especially, are often
carried out interactively so that the effects
of each particular operation can be evaluated
by the user. This is because it can be very
difficult to set conditions so that features
clearly seen by eye are separated unambiguously
by an image analyser using a digitised image.
For instance, the eye is very sensitive to edge
information (interpreting the edges as blacker
than they really are) and not to absolute
intensities, whilst the reverse applies to
image analysers (the absolute intensity
associated with each pixel forms the basis for
feature identification. Thus while the brain
uses a number of criteria simultaneously to
differentiate regions in an image, an
instrument (image analyser) is much more
restricted and compromises may have to be made
by a user.

4.1 Image correction and calibration

Two major corrections that may need to be made
to an image are firstly for geometrical
inaccuracies arising from, say, a camera lens
and video camera and secondly for non-
uniformity in the illumination. In addition,
random noise must be minimised and the system
itself must be calibrated in real units so that
features analysed have dimensions which are
related to the original object.

13

The first correction has been covered in section 2.1 above; distortions in microscope lenses and in cameras can be allowed for by storing reference grids and then using these to adjust all subsequent images. With regard to the second correction, if the original lighting of an image is not uniform then the same features within an image may have different intensities depending upon their location in the image and so some features might be detected and others not. Again, this effect can be minimised by the collection of a reference or "background" (e.g. a featureless plain white object) image which will show any uneven illumination or "shading"; this reference image can be used to "background correct" subsequent images and so help compensate for illumination variations.

Random noise can be minimised easily if the original object being viewed is static since a number of separate images can be collected and averaged to provide the digitised image in the frame store; the premise is that random noise is unlikely to affect the same pixel location twice. The improvement in signal to noise ratio changes by the square root of the number of images averaged; generally, there is little to be gained by averaging more than 16 images and, in many cases, the averaging of 8 images is sufficient. Note that if the image analyser is in a location where the object is affected by vibrations, such averaging will result in a worse result together with a blurred final image.

Finally, calibration of an image is essential if quantitative data in real units are required. Typically, with the same experimental arrangement (lens and magnification) as will be used to collect the other images, an image is collected of a known length, e.g. using a ruler, graticule, microscope grid or micron marker. This image is then displayed and is used to calibrate directly the size of individual pixels.

4.2 Image enhancement and analysis

Having collected and corrected an image, the measurement of data · can be undertaken. This can be carried out in two ways.

In the first, absolute intensities from optical density measurements are used directly; these have been used traditionally in biological fields and now increasingly in transmitted light studies in materials science. In such cases, it is particularly important to calibrate the grey scale to provide absolute values, to ensure there is a linear response to the incoming light, and to minimise drift and noise in the entire system.

In materials' science, absolute intensities are not as widely used as the approaches embodied in the second approach. In this, it is the selection of features that is particularly important and the precise grey scale values are used only to identify and separate features. Thus an image comprising many grey scales ultimately can be reduced to a "binary" representation in which all selected or thresholded items are represented by a value of 1 and the rest of the image is regarded as background with a value of 0.

In this section, the means by which the quality of an image can be assessed (use of a grey level histogram) and by which features can be selected (segmentation) are presented.

(a) Grey level histogram

The numbers of pixels having particular intensities or grey levels can be displayed as a histogram for the entire range of grey level values e.g. 0 to 255. If an image contained only very dark (values near 0) and very light (values near 255) features, the corresponding histogram would be as shown in Fig. 4a, whereas if the image varied uniformly from black to white then a histogram spanning the entire range of intensities, as Fig. 4b, would be obtained. In practice, the histogram of an image will reflect the variations in the numbers of dark and light features (e.g. see Fig. 4c) and also the extent to which there is

15

good crisp delineation between each of the phases.

If the illumination has been selected well and there is a full range of grey levels in the object, the histogram will span the entire range of intensities, as was shown in Fig. 4c. However, if there is insufficient signal or too great a signal (leading to saturation) then histograms such as those in Figs. 4d and 4e will be observed. In such cases, changes in the conditions might be made to improve the distribution and so facilitate the subsequent analyses. If this is not possible, various operations (redistributing the grey scale values) can be carried out at this stage to optimise the intensity histogram, e.g. grey level histogram equalisisation tends to flatten the grey level histogram with a corresponding improvement in contrast.

The histogram may also be useful as the first step in evaluating whether or not two phases, say, in an image will be easy to identify, see Fig. 4c and "thresholding" below.

(b)   Segmentation
Having obtained an image with a suitable range of intensities, it is necessary to select the features that are to be measured or analysed; this process of separating objects in an image from their background is generally called *segmentation*. There are three main ways this can be accomplished:

-   subtract entire images from each other to leave any differences as the only features in the resulting image (an approach which is particularly successful in applications involving quality control and security);

-   select groups of pixels with similar properties e.g. using absolute intensities of different parts of the image which is often called "thresholding";

16

- identify boundaries which have high rates of change of intensities, called "edge detection".

The latter two approaches are considered below in more detail. In terms of pixels, these two approaches can also referred to as "point-dependent" and "neighbourhood" techniques respectively.

Thresholding is the setting of a window based on a range of intensity values (or grey scales) such that only those features whose intensities lies within the selected range will be included in any subsequent analyses. If the histogram of an image shows clearly distingushable groups of intensities (e.g. as in Fig. 4c) then thresholding is likely to be successful in separating the features corresponding to different intensity. This is a major way by which features are identified in image analysis. However, it should be noted that in practice, even having taken care in specimen preparation and in selecting conditions, images seldom show clearly distinguishable groups and it can be difficult to threshold only features of interest.

The question of what thresholds to select is crucial to the accuracy of the subsequent analyses. Idealised video signals from large and small particles of two different phases are shown in Fig. 5a together with possible threshold levels (after Exner and Hougardy [4]). All particles would be detected if threshold window AC was chosen while only the darker particles would be selected if threshold window BC was used. In practice, however, the video signals from the cameras do not display such true square wave behaviour and more realistic video signals for the same four objects are shown in Fig. 5b. Two effects can be seen.

(i) Particles of the same phase (hence theoretically of the same original intensities) do not result in the same magnitude video output voltage. Clearly,

17

if a threshold window BC was chosen, the
small dark particles would be excluded.
If, to compensate for this and to include
these small particles, the threshold window
was increased to AC, say, then the larger
particles of the lighter second phase would
be included as well. Hence it would not be
possible in this instance to include all of
the first phase by simple thresholding.
Note that the overall particle size is not
as important in this example as the
orientation with respect to the scan line
of the camera since this influences the
response of the video voltage.

(ii) The detected sizes of particles (shown by
the lengths of various parts of the scan
lines in Fig. 5) vary according to the
magnitude of the threshold selected. Thus
the same particle will appear to have
different sizes depending on the threshold
voltage used. Although a threshold value
may be found to measure accurately the size
of, say, the large darker particle, the
size of the smaller particle (and those of
the other phase) will not be correct.
Image enhancement can be used to reduce
this effect; for instance, differentiation
of the video signal and then adding the
differentiated signal to the original video
signal can make the edges of the objects
steeper or "sharper". However, for
accurate results, it is imperative to use
controls and/or a standard operating
proceedure that will help to ensure errors
are systematic rather than random.

Edge detection relies on accentuating those
parts of an image where the intensity is
changing rapidly, such as at the boundaries
between two phases. This is an example of
image enhancement; the original values of
intensity for certain pixels will be changed to
facilitate subsequent identification of
features, e.g. by seqmentation. In practice,
edge detection can be achieved using digital

18

filters to approximate mathematical procedures such as differentiation. The intensities of each pixel in the original input image are treated using a "mask" or "kernal" such that the corresponding output intensity of each pixel is given a new value which depends not only on its own original value but also in some way on the intensities of a defined number of nearest neighbours.

Consider the example above of edge sharpening by differentiation. A number of filters exist for differentiation including an isotropic high pass filter, the mask for which is shown below.

-1	-1	-1
-1	9	-1
-1	-1	-1

Each input pixel is adjusted according to:
  [its own value multiplied by a factor of 9]
from which is subtracted
  [the intensity multiplied by 1 of each of its
    adjacent eight nearest neighbours].
Hence in the output image, a high value of intensity will arise wherever the intensity in the original input image changed rapidly, and a low or zero value will arise wherever the original intensity was uniform from pixel to pixel. Thus features should be well distingushed from each other and so should be easier to threshold (schematically illustrated in Fig. 6).

(c) Enhancement

An example of enhancement has been presented above, namely the use of masks or kernals to sharpen edges. Some other approaches include:

  smoothing ("low pass") filters - each pixel
    is replaced by the average grey scale
    value of itself and its neighbouring
    pixels, and hence is similar to

19

integration. This decreases noise (rapid or high-frequency changes in intensities) but will, therefore, lead to a loss in details at edges (where intensities also change rapidly) and a defocussed appearance of the image.

median filters - similar to smoothing but maintain edge details.

Laplacian - good for detecting edges but flattens overall contrast.

high pass filters - remove low frequency changes in intensity across an image (e.g. gradual shading changes due to uneven illumination) while variations associated with image features are relatively unaffected. [Note that the background subtract approach to correct illumination - see section 4.1 - is generally more satisfactory.]

gamma correction - increases the available intensity values (hence contrast) at one end of the grey scale at the expense of those at the other (which are suppressed). This is a technique familiar to scanning electron microscopists.

It is evident that images may be enhanced in many different ways using filters such as those above and for different reasons. However, it is possible that, having successfully segmented an image, there still may be features that need to be separated (e.g. to isolate contacting particles), or that need to be merged (e.g. fibres split as a consequence of noise having isolated segments). Hence there are other operators available which can be used automatically on thresholded features and prior to the measurement of features. However, use of these operators is dependent of the user making assumptions about the sizes or shapes of the features to be measured and so may be used unwittingly to reinforce incorrect assumptions. The following are examples of such operators:

erode/shrink   - remove one layer of pixels from perimeter of feature;

dilate/swell   - add one layer of pixels to perimeter of feature;

skeletonise   - reduce feature to a single line of pixels;

skeletonise   - reduce width of background to
background   that of a single pixel (e.g. useful to reduce the boundary regions between thresholded grains prior to measurements of grain size);

circle   - separates touching convex
separation   objects e.g. by joining any two concavities in a feature.

These operators can also be modified; for instance, 'erode without splitting' would mean that a feature that would have been separated by the erosion remains as one unsplit object.

In addition to these automatic operations, it is possible to edit images manually. By means of a light pen or mouse, features can be isolated, separated, included, excluded, expanded or reduced in size. Clearly, such operations are both very subjective and also time-consuming. Manual editing should be used with caution and, wherever possible, avoided.

In summary, as was stated in the introduction, image enhancement techniques may result in changes to the original input data and hence can introduce errors in subsequent measurements; accordingly, image enhancement must be used only when absolutely necessary and with suitable care. In recent years, there has been considerable development of the algorithms that can be used both to distingush and to enhance features. Accordingly, caution must be exercised in selecting the right approach for a particular problem; it is essential to set up model situations in order to simulate the problem to be investigated and thus ensure that the approach adopted is both valid and accurate.

5. MEASUREMENTS WITHIN IMAGES

5.1 Feature and field measurements

Having resolved an image into features to be measured, the subsequent measurements are normally quite straightforward. There are two main approaches. *Feature* measurements can be made of individual objects (contiguous groups of pixels) and include measurements of individual areas, perimeters, lengths. *Field* measurements are made with reference to an image area defined using an adjustable "frame", i.e. it is the relative numbers of pixels with defined intensities within the frame area that are important (e.g. the relative amounts of each phase in a two-phase mixture).

Typical *primary* feature parameters include perimeter, area and various lengths or distances. These latter are often based on Feret diameters, otherwise known as caliper lengths; the equivalent of a caliper is placed around the feature to be measured and the resulting length is a Feret diameter, see Fig. 7. Since a feature is unlikely to be a perfect circle (as imaged in two dimensions), the actual value of a Feret diameter will vary with orientation of the caliper. Accordingly, many systems take a number of Feret diameters at 5 or 10 degree rotations around the feature; then length is taken as the longest, breadth as the shortest Feret, and diameter as the average of the 36 or 18 Ferets measured around the feature. The use of a large number of Ferets means that the primary parameters generally are not significantly affected by the orientation of the object in the field of view.

22

Clearly, correct interpretation and use of the primary parameters is essential. For instance, if a rectangular object is measured, the shortest Feret would correspond to the width but the longest Feret would not be the rectangle's length but would be its diagonal (see Fig. 8). The actual length could not be measured accurately in a simple way but is easily derived, e.g. by dividing the measured area by the shortest Feret (the width). The perimeter of a thresholded object will not necessarily include the perimeter of any internal hole, and even concave indentations on the surface may be inaccurately measured depending on the algorithm used.

Many image analysers allow the direct calculation of *secondary* parameters from the primary parameters; the length above would be a simple example of such a parameter. There are many expressions for secondary parameters and the usual care must be taken in both selecting and then checking (with model data) the validity of an expression for a particular application. As an example, consider measuring the length and width of curved fibres (increasingly important with regard to the development of fibre composites). If primary paramters are considered, considerable inaccuracies will result, the natural consequence of putting a curved object into a caliper (the lengths will be too short and the widths too long). Instead, the length l and breadth b of a curved fibre can be expressed as:

$$l = 0.25 \ [\ P + (P^2 - 16A)^{0.5} \ ] \quad \text{and}$$

$$b = 0.25 \ [\ P - (P^2 - 16A)^{0.5} \ ]$$

where P is the perimeter and A the area of the fibre.

5.2 Shape factors

A shape factor is a dimensionless expression which can be used to classify thresholded features merely on their shapes and regardless of their sizes. For instance, consider the expression $P^2/4\pi A$ which can be used to assess roundness. If a feature is perfectly round *and* smooth, then the expression will have a value of unity and will deviate from this value if the feature is either not round or not smooth. This can be seen in Fig. 2 which also serves to reinforce the point made in section 3.2 regarding magnification. The gear wheel would be classified as round using low magnification (shape factor of 1.04) yet the *same* object would not be classified as round if viewed at a higher magnification (shape factor of 1.41). Accordingly, a better shape factor in this case might be $\pi l^2/4A$ (where l is the average Feret diameter) which again is unity for a circle but, unlike the first expression, is independent of roughness; the gear wheel in Fig. 2 will maintain a value of unity at all magnifications.

Another common shape factor is aspect ratio (length divided by breadth). This would clearly separate elongated inclusions from others. However, if used for curved fibres, the derived values for length and breadth (see above) would have to be used.

A recent development in shape factors is the use of fractals and this concept may eventually be used routinely to get around existing ambiguities such as indicated for the gear wheel example above, as well as to describe the shapes of objects generally. Consider imaging an object at different resolutions. If the perimeter looks more "crinkled" with increased resolution (smaller measurement steps around the object), there can be a uniform increase in measured perimeter due to more detail being observed while the area remains constant. Thus it is not possible to state precisely what the value of the perimeter should be. However, it is possible to describe the fractional dimension or *fractal* by which

24

the perimeter increases as resolution increases
(i.e. as the measurement steps around the
perimeter decrease). The example often quoted
is the change in length of measured coast line
as the length of the measuring stick decreases.
The use of fractals is not confined to lines
and fractals have values between 1 and 2 for
boundaries and between 2 and 3 for surfaces.
Their appeal is that characteristic shapes can
be described by certain fractal values [5].
While at present, their use in image analysis
is not extensive, it is envisaged that this
will change as they become better understood
and they will form an important part of
stereology.
From the above, it is evident that shape
factors can be extremely useful but, as
illustrated by the simple example of the gear
wheel, it is clear that caution must be
exercised in their selection and usage.

5.3 Discrimination by size or shape

From the brief summary of segmentation (section
5.1), it is obvious that features within images
cannot always be separated merely by
differences in intensities. The use of size
and/or shape discrimination can aid the
selection of features. Having thresholded the
image, a set of conditions based on size (using
one or more of the primary parameters) or shape
(using expressions similar to those in section
5.2) can be specified by the user to include or
exclude features from the results being
gathered.

5.4 Statistics

The measurements made of features within an
image (or number of images) will be stored for
subsequent interpretation or display. Clearly,
it is imperative that sufficient data are
collected from representative images. This can
be very tedious if manual methods are used and
completely automated systems (including
stepping of specimen stages, auto-focussing,

intensity control) are advantageous providing that sufficiently reliable feedback signals are included to ensure that the data are collected under appropriately controlled conditions. No attempt will be made to discuss the criteria by which reliability can be assumed; different problems require different approaches and there are many sources of statistical tests and tables for the design and execution of scientific experiments [6]. However, as an example, it is instructive to note that at least 2000 measurements are required if features are to be classified by size into twenty groups with an acceptable degree of confidence [7].

It is essential that the statistical aspects of data collection and processing are considered prior to collecting data. The selection and preparation of the specimen, choice of magnification, the number of features within an image and the number of images measured will all influence the subsequent accuracy and reliability.

6. STEREOLOGY

A problem in characterising three-dimensional microstructures using image analysis is that data are obtained from two-dimensional images. If the results are to be correlated with properties (naturally representative of a three-dimensional microstructure), then the two dimensional data need to be transformed into a three-dimensional representation. Stereology is the means by which this is achieved; sets of mathematical relationships are used to relate the two-dimensional data to the original microstructure in three-dimensions [8].

Stereology is, therefore, one of the most important steps in image analysis. In recent years, image analysers have become more widespread and have been used in increasingly diverse applications due to their improved ease of use, greater resolution and accuracy, and reduced costs. Ironically, the appreciation and usage of stereology does not appear to have increased correspondingly. In fact, many of the fundamentals of stereology were proposed, derived and tested using other older image analysis techniques such as eye-piece grids in light microscopes and automated point counters.

There are a number of fundamental relationships which are independent of the features being analysed and hence have widespread usage. For reflected light microscopy, these are:

$$V_V = A_A = L_L = P_P \qquad (mm^0)$$

$$S_V = (4/\pi)\ L_A = 2P_L \qquad (mm^{-1})$$

$$L_V = 2P_A \qquad (mm^{-2})$$

where:

> V_V, A_A, L_L and P_P are the volume, area, line and point fractions respectively;

> S_V is the surface area in a unit volume;

> L_A is the line length or perimeter in a unit area of view;

> P_L is the number of points lying on a unit length of line;

> L_V is the line length or perimeter of features in a unit volume;

> P_A is the number of point features per unit area of view.

Variants of the above equations exist for transmission light and electron microscopies in which, for example, overlaps of features may occur (some features will lie in different planes in the z-direction in the field of view and so will overlap and obscure others).

The use of the above simple stereological transformations enables the two-dimensional data collected by an image analyser to be transformed readily to some three-dimensional structural information. However, properties such as particle size distributions, mean particle size and numbers of particles per unit volume cannot be obtained using these simple transformations. For such properties, more precise knowledge of particle shape is required and needs to be either measured (e.g. by analysing successive serial sections from a specimen - facilitated in metal samples by the use of hardness indents to locate the same area after polishing and also to estimate from their decreased area the amount of metal removed during polishing) or assumed (e.g. by modelling the particles as simple regular shapes such as spheres, ellipsoids, dodecahedrons). If particle shape is measured experimentally or assumed, then expressions for particles can be used. For instance, assuming spherical particles, the mean particle size (D_m) is given by:

$$D_m = (2/\pi) \ P_L/N_A \qquad \text{for individual particles}$$

$$\text{or} \quad D_m = 3/(2P_L) \qquad \text{for grains (filling the entire volume).}$$

Naturally, the assumption of simple regular particle shape does not precisely describe real materials in which there will be statistical distributions of more irregular shapes. The extent to which a distribution deviates from an assumed shape strongly affects the accuracy of any stereological result. Nonetheless, the use of stereological transformations, following a considered experimental approach (e.g. using serial section) allows accurate descriptions of real microstructural features such as

- oriented or directional microstructures (phase distributions, dislocation arrangements);

28

- grain, fibre and particle sizes, shapes and distributions;
  - contiguity (interconnectedness - of relevance in nucleation and growth studies).

7. SUMMARY

The aim of this chapter has been to introduce briefly the topic of image analysis; the terms used, the various stages involved and possible pitfalls. With regard to pitfalls, there is a real danger that, because of their apparent ease of use and user-friendliness, the numbers generated by techniques such as image analysis are "correct" and are to be believed. As has been discussed, considerable care must always be exercised when setting up systems to gather data and subsequently both in providing appropriate checks on the data obtained and in their analyses.

Like many other rapidly developing numerical techniques, image analysis is realising its considerable potential. Using stereological transformations, it is becoming easier to relate quantitative microstructural data to compositions, fabrication histories, physical and mechanical properties for materials as diverse as metals, ceramics, glasses, polymers, wood, concrete and composites. Hence, image analysis is helping in the modelling of materials' behaviour and in providing the basis for subsequent predictions of expected service performance.

8. REFERENCES

8.1 Specific references

1. R. VAN WEZEL: 'Video Handbook', 2nd edn, 1987, London (England), Heinemann.

2. R.C. COCHRANE: 'Optical microscopy",pp.43-93 in reference 3.

3. E. METCALFE (ed.): 'Microstructural characterisation', 1988, London (England), The Institute of Metals.

4. H.E. EXNER and H.P. HOUGARDY: 'Quantitative image analysis of microstructures', 1988, Oberursel (Germany), Deutsche Gesellschaft für Metallkunde Informationsgesellschaft mbH.

5. J.C. RUSS, 'Fractals', pp.169-175 in R.W. CAHN (ed.): 'Encyclopaedia of Mat. Sci. and Engineering', Suppl. Vol. 1,1988, Oxford (England), Pergamon Press.

6. D.L. DAVIES and P.L. GOLDSMITH: 'Statistical methods in research and production',4th edn, 1972, Edinburgh (Scotland), Oliver and Boyd.

7. H.E. EXNER: *Int. Met. Rev.*,1972,**17**,25.

8. R.T. DeHOFF, 'Stereology,' pp. 4633-4637 in M.B. BEVER (ed.): 'Encyclopaedia of Mat. Sci. and Engineering', Vol. 6, 1986, Oxford (England), Pergamon Press.

8.2 General references

R.T. DeHOFF and F.N. RHINES (eds.): 'Quantitative metallography', 1968, New York (U.S.A.), McGraw-Hill.

JOYCE LOEBL: 'Image analysis: principles and practice', 1985, Exeter (England), Short Run Press.

F.B. PICKERING: 'The basis of quantitative metallography, 1976, London (England), The Institution of Metallurgists.

J.C. RUSS: 'Practical stereology', 1986, New York (U.S.A.), Plenum.

E.E. UNDERWOOD, 'Quantitative metallography',
pp. 123-134 in Metals Handbook, 9th edn, Vol.9,
'Metallography and micro-structures', 1985,
Ohio (U.S.A.), American Society for Metals.

8.2 Specialised topics

G.A. BAXES: 'Digital image processing - a
practical primer', 1984, New Jersey (U.S.A.),
Prentice-Hall.

K.R. CASTLEMAN: 'Digital image processing',
1979, New Jersey (U.S.A.), Prentice-Hall.

R.C. GONZALEZ and P. WINTZ: 'Digital image
processing', 2nd edn, 1987, Massachusetts,
(U.S.A.), Addison-Wesley.

W.K. PRATT: 'Digital image processing', 1978,
New York (U.S.A.), John Wiley.

A. ROSENFELD and A.C. KAK: 'Digital picture
processing', Vols 1 and 2, second edn, 1982,
New York (U.S.A.), Academic Press.

E.E. UNDERWOOD: 'Quantitative stereology',
1970, Massachusetts (U.S.A.), Addison-Wesley.

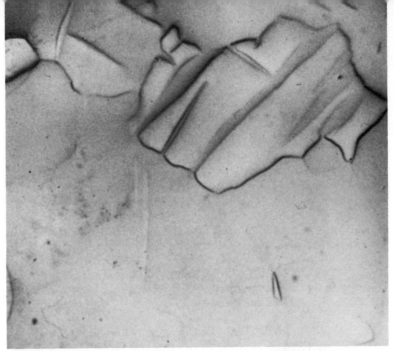

Fig.1. Light micrographs of zinc:
        (a) standard illumination;
        (b) polarised light;
        (c) as (b) but sample rotated.

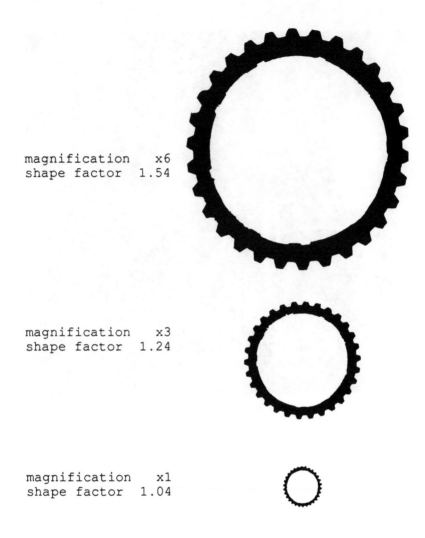

magnification x6
shape factor 1.54

magnification x3
shape factor 1.24

magnification x1
shape factor 1.04

Fig.2. Profiles of gear wheel at different
magnifications. Using a shape factor
of $P^2/4\pi A$ (see text), the gear would be
classified as round only at low
magnification.

Fig. 3(a)

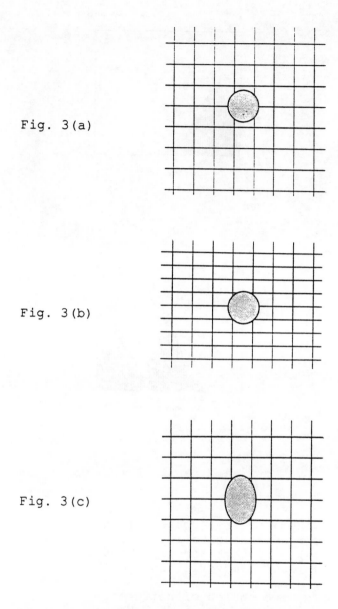

Fig. 3(b)

Fig. 3(c)

Fig. 3. Schematic SEM micrographs showing
        sphere on grid:
            (a) viewed normal to grid surface;
            (b) viewed with 45° tilt;
            (c) as (b) but with tilt correction.

Fig. 4d

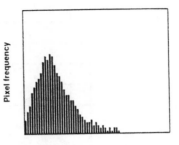

Grey scale values

Fig. 4e

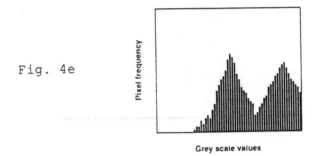

Grey scale values

Fig. 4. Schematic grey scale histograms
corresponding to images with:
(a) only dark and light features;
(b) uniform variation from black to
white;
(c) two phases of different
intensities;
(d) insufficient lighting (hence
signal);
(e) excessive lighting (hence
saturation).

Fig. 4a

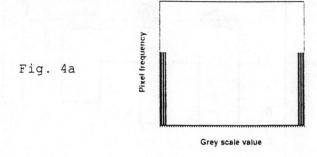

Grey scale value

Fig. 4b

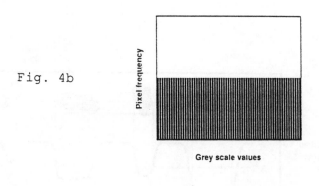

Grey scale values

Fig. 4c

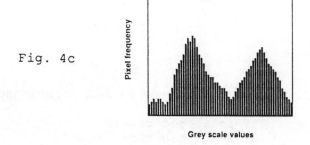

Grey scale values

37

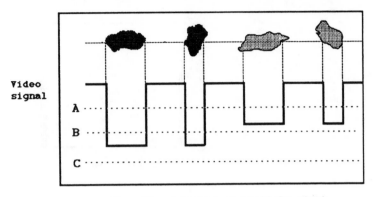

Video signal
A
B
C
Scan time (measure of particle size)

Fig. 5a

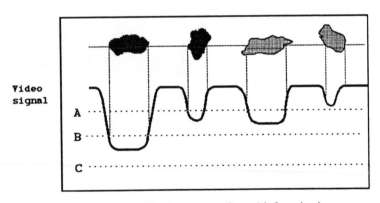

Video signal
A
B
C
Scan time (measure of particle size)

Fig. 5b

Fig. 5. Video signals from camera corresponding to:
(a) idealised case;
(b) realistic case.
[After Exner and Hougardy [4]]

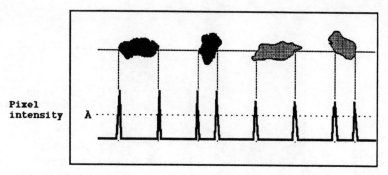

Line scan across sample

Fig. 6. Intensity variation after
differentiation.

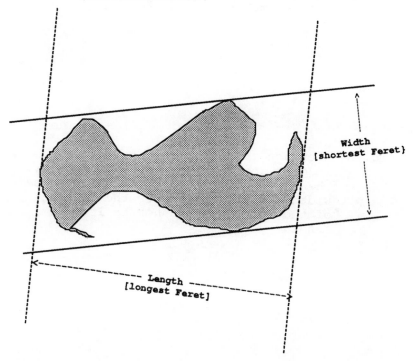

Fig. 7. Representation of longest and
shortest Feret diameters. Note that
the angle between these two diameters
will be governed by the sample shape.

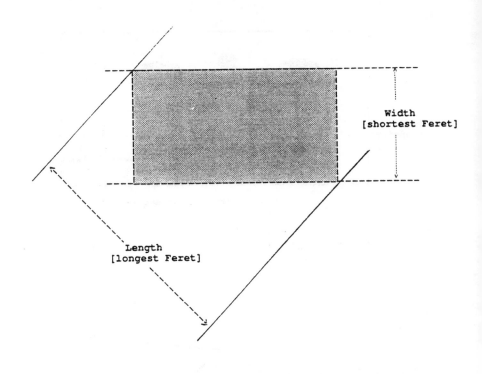

Fig. 8. Rectangular object measured by the
longest and shortest Feret diameters.
The shortest Feret corresponds to the
width and the longest to the diagonal
of the rectangle and not to its
length.

2:The use of databanks in materials science

K WILLIAMSON

UKAEA, Warrington

1   INTRODUCTION

The vast majority of the world's computers are used either wholly or primarily for the storage and retrieval of data. In this respect the term "computer" is perhaps a misnomer, since most of the operations carried out by such computer systems involve little or no computation. The "data explosion" has been one of the most important trends of the latter quarter of the present century, and this has been facilitated by the ability of computer systems to store large amounts of data, and then to retrieve and present required selections from the data much quicker than could be achieved by other methods.

Many databanks are stored on large mainframe computers, and involve fairly trivial operations on huge amounts of data. Some obvious examples are the Driver and Vehicle Licensing system at Swansea, which keeps track of, and issues licences to, the millions of cars and drivers in the country. This involves thousands of repetitive operations every day, which requires a powerful system but does not place any great demands on it other than speed.

Other systems have equally large storage requirements, but also demand extensive processing or analysis of data. An example of this is the police fingerprint analysis system, which is intended to match an unknown fingerprint to the stored digital representations of known prints. This involves very complex pattern matching processes as well as large amounts of storage.

These two applications are typical of large databanks
which need powerful mainframe computers, highly skilled
operators and, often, teams of programmers to keep the
system going. The need for absolute security of data, in
both meanings of the term, mean that the data must be
protected by the use of passwords, and secured against
loss by careful backing up and archiving. However, the
availability in recent years of powerful personal
computers with adequate data storage, and suitable
database and other data handling software, means that
quite large databanks may be set up on a desktop.

This paper is intended as an introduction to the use of
small computer systems for the storage of an
organisation's own data, and also to the availability
and usefulness of larger public data sources.

2   MATERIALS DATA REQUIREMENTS

The requirement for up-to-date, valid and reliable data
on the wide range of materials in use in the modern
world is of quite paramount importance. All designers
and structural engineers need to know about the
materials they use in intimate detail, both to ensure
the safety of the structures they design or work with
and to optimise the efficiency of plant or equipment by
operating them under ideal conditions.

The provision of such data has occupied manufacturers,
research organisations and large users since the start
of the industrial revolution. Most data has historically
been in the form of printed tables of results, or graphs
of actual or idealised data.

While such data sources have been, and will continue to
be, of great value, the recent advances in computerised
data storage systems have allowed their speed and power
to be used to make materials data much more readily
available to users. However, the detailed data
requirements placed on a materials data system are much
more complex than for most data systems, and this is
reflected in the apparent lack of progress in some
areas.

Some of the advantages of computerised data storage, and the associated problems, are discussed in the following pages.

3   DEFINITION OF A DATABANK

A databank is a source of available, structured data. The use of the word databank rather than database is deliberate. The former stresses the information included in the system, while the latter is a more restrictive term describing one particular form of computer-based information source.

An important point is that a databank need not be on a computer. Any available source of categorised information may be called a databank. This may apply to reference books, videotapes and films, all of which can store information for future retrieval. The most common non-computerised databank, and one on which much of the terminology and techniques of computerised data storage are based, is the traditional manual filing system. However, the present study will deal with computer systems, although with some reference to manual systems where a comparison may help to illustrate a point.

Even within computer systems, databanks or data sources are not restricted to the conventional database. Any stored data may be called a databank. This is not a trivial point; much of the usage of computers in science and engineering is concerned with the storage, manipulation and retrieval of data, often by means of software systems which would not be called database systems, such as spreadsheets and statistical programmes. The important thing here is the organisation of the data and its availability to the user, rather than the software used to manipulate it.

With all this in mind, a definition of a useful databank covers the following points:

a)   **Retrievability**. Data should be stored in a form which allows for convenient and rapid access.

b)   **Selectability**. It should be possible to select data based on a range of criteria on one or more items of

43

information. For instance, in a mechanical properties databank, a steel with U.T.S. above 450 M.Pa. and ductility greater than 50%, and also with between 0.1 and 0.2% of carbon, may be required.

c) **Sortability**. Data may be required to be ordered in different ways. For instance, in a databank of constructional steels, arranging the data in ascending order of U.T.S. will enable the strongest materials to be quickly picked out. Where the requirements are more complex, for instance if high U.T.S. and high yield strength are both important, the data may be arranged according to both criteria, and again the most suitable materials may then be seen at the upper end of the range.

In addition to sources of data not being restricted to database systems, the converse is also true; database systems are not only useful for storing and retrieving data in a standard form. The power and flexibility of modern systems are commonly used to manipulate and analyse the stored data, either within the database system or by feeding the selected data to a separate data analysis system. This can be either for repetitive calculations, for instance, conversion of peak stress to peak strain values using the elastic modulus, or for regression analysis to give best-fit curves which may then be used to predict unknown data.

4   APPLICATIONS OF DATABANKS

Computerised databanks generally fall into two categories.

a)   **PRIVATE** databanks, which are set up within an organisation for the maintenance of their own data.

b)   **PUBLIC** databanks, which are set up for access by authorised users outside the generating organisation, normally as a commercial venture.

The two types of databank may be similar in structure, and may use similar hardware and software. The important difference is in the intended use of the information rather than the technical details of the database system.

Having made this point, however, it is normal to put extra emphasis on the database system technology concerned in private databanks, in which the user may well be involved in setting up the system. With public systems, more emphasis is placed on the method of access, the cost and the usefulness of the data included.

5   TERMINOLOGY

There are certain universal terms used in the context of databanks. All are exactly analagous to the card-index file example, and an understanding of them is important in the subsequent discussion.

First are the terms used to describe the structure of the data in the databank. They are FILE, RECORD and FIELD. Taking them in reverse order:

a)   **Field**. A field is one piece of information in the databank. It is the smallest unit of data which the databank normally handles, and is the unit used in selection, sorting and indexing of the data. It is equivalent to one item on a record card, for instance the surname in a personnel file.

b)   **Record**. A record is an assembly of fields, all relating to the same larger unit. It is the equivalent of a single record card in a card index file.

c)   **File**. A file is an assembly of records, and with many small databanks is the largest unit of data, i.e. it is the databank. With larger banks, however, the preferred structure is to have multiple independent files, with links between them through equivalent fields.

Next are some terms describing operations carried out on the databank; SORT and INDEX.

d)   **Sorting** is fairly self-explanatory. The records in the databank are rearranged according to the magnitude of one or more data fields. Sorting is not normally necessary with modern systems, since its effect can be duplicated by indexing.

e) **Indexing** produces the same effect as sorting, but without altering the actual order of the records in the databank. Instead a separate INDEX FILE is created which lists the record numbers in sorted order. Any subsequent operations which are required in indexed order look up the correct order in the index file.

Other terms are to do with the method of operation of the database system, and the internal ordering of the data within the system.

f) **Relational** and **Hierarchical**. Traditionally, database systems involving multiple files have used a hierarchical structure, with fixed linkages between files set up as part of the file structure. A simple analogy for this is a family tree, with the "relations" between members of a family fixed immutably at birth.

Most modern database systems claim to be relational, with no fixed links between files. The structure of the database is imposed on the data by the overlying system, thus allowing complete flexibility in the the provision of links between, and the retrieval of, data. The final result can appear hierarchical, as in the simple example quoted in this paper, but the advantage of a relational system is that the apparent structure can be changed without changing the data files in any way.

There is actually much more to a relational system than this, and a set of guidelines has been published by Codd (1, 2, 3) to determine how well commercial database systems fit the relational model.

6   ADVANTAGES OF COMPUTERISED DATABANKS

Computerised systems do have definite advantages over other forms of data storage; this is why the majority of computer systems in the world are used wholly or partly for database applications.

Four of the main advantages are:

a) **Classification**: the indexing, selection and sorting
capabilities of computers make the classification of
disparate data quick and convenient.

b) **Validation**: evaluation and validation of stored
data is crucial to the usefulness of the databank.
Computerised systems aid greatly in this important task.

c) **Analysis**: Raw data may be subjected to statistical
or other numeric analysis, to produce models or
equations which are directly useable to solve real
problems.

d) **Communication**: Computer-stored data may be
transferred or via communications channels and networks,
to greatly facilitate the dissemination of information.
This may be considered as the other side of the coin
from the much-publicised (and often over-emphasised)
problem of "hacking" or illegal access to computerised
data: it is the very accessibility of the data which
produces the problem.

7   USES OF COMPUTERISED DATABANKS

All industrial processes require materials, and many
modern processes subject their materials to extremes of
pressure, temperature and/or corrosive environment. Some
obvious examples are the aeroplane and aerospace
industry, nuclear power and oil exploration, but many
chemical processes also exploit materials to the utmost.
The understandable wish to increase the efficiency of
all these processes requires constantly increasing
temperature, pressure, etc., which can result in
constantly reducing safety limits. The availability of
valid, relevant and up to date data is vital in such
circumstances.

Some typical uses of stored data in such applications
are:

a)   Provision of conveniently classified data on common
materials.

b)   Development of design codes and constitutive equations to improve the efficiency of utilisation of, and safety limits for, structural materials.

c)   Substitution of alternative materials for scarce or expensive existing materials.

d)   Possible future integration of materials data into CAE (computer-aided engineering) systems.

8   PROBLEMS WITH COMPUTERISED DATABANKS

The advantages of computerised data in the materials field are clear cut. Unfortunately there are problems in obtaining valid data and in matching the stored data to the requirements of users.

8.1 Validation of data

Data intended for incorporation into a general databank is often from many different sources, which are not necessarily compatible. The data may be described differently, or be produced using different test methods, or be differently evaluated from the raw data. Moreover, data may be presented in various forms; for instance, as numeric data from one source and graphical from another.

All this diverse data must be validated and checked for compatibility before being incorporated in the databank. This adds greatly to the costs and complication of setting up such a data source.

8.2 Data output requirements

The materials data requirements of different data users can vary greatly. For instance, design engineers concerned with producing safe designs require validated lower limit or worst-case mechanical properties data, often incorporating safety limits, while structural engineers wishing to analyse the true stress distribution in a structure require numerically evaluated data models, and materials scientists wishing

48

for comparisons with their own data require raw data.

For a public databank to be fully useful, all these requirements must be met. This places an onus on the flexibility of data retrieval in the system.

Different forms of data output are useful in all cases; for instance, graphical representation of a design curve can be far more meaningful than lists of numbers. Inbuilt data analysis can also be very useful for some users.

9   AN EXAMPLE DATABASE

In order to illustrate some of the concepts of a working "private" database system, I shall show the process of building up a database which will later be used for the selection of a structural steel for a particular component.

The complete database will consist of a number of separate files, each concerned with a different aspect of materials behaviour, but, very importantly, all interlinked.

The first file, which we shall call **STEELS,** will contain the names and sources of a range of structural steels. All this information will most probably be found in the steel manufacturers' literature.

The second file, called **PHYSPROP,** will include basic physical properties, such as thermal expansion coefficient and density.

The third file, called **HEATREAT,** will contain any heat treatments given to the material before delivery, and any required before use.

The fourth file, called **CHEMPROP,** will contain the typical chemical constitution and allowable range for each constituent of the steel.

The fifth file, called **MECHPROP,** will contain the standard mechanical properties of the steels, such as U.T.S., yield strength and tensile ductility.

For the sake of simplicity, we shall leave out any more complex mechanical properties such as creep and fatigue strength.

The actual contents (the fields) of each file will be as follows:

STEELS: Name of steel
             Steel type (A.I.S.I. or B.S. ref. no.)
             Manufacturer
             Price code
             Availability

PHYSPROP: Name of steel
             Steel type (A.I.S.I. or B.S. ref. no.)
             Thermal expansion coefficient
             Density

HEATREAT: Name of steel
             Steel type (A.I.S.I. or B.S. ref. no.)
             Heat treatment before delivery: temperature
                                               time
                                               environment
             Heat treatment required: temperature
                                               time
                                               environment

CHEMPROP: Name of steel
             Steel type (A.I.S.I. or B.S. ref. no.)
             Batch ref. no.
             % Cr - typical
             % Cr - minimum
             % Cr - maximum
             % Ni - typical
             % Ni - minimum
             % Ni - maximum
             etc. for each constituent

MECHPROP: Name of steel
             Steel type (A.I.S.I. or B.S. ref. no.)
             Batch ref. no.
             U.T.S. at room temperature
             Elastic modulus
             0.2% yield strength
             Tensile ductility

Note that each file contains two common fields; the name
of the steel and its A.I.S.I. or B.S. reference number.
These are the **linking fields** between the separate files,
and are the key to the structure of the whole database.

9.1 The database structure

The structure of the database may be considered as five
interlinked files, each containing specific information
about the steel. All are linked or related to each other
by the two fields mentioned above. However, some files
deal with general properties which are the same for any
batch of the steel, while others contain properties,
such as mechanical properties, which may be different
for different batches or casts. Those files which deal
with batch-specific information use the batch number as
a second linking field.

The reason for the multi-file structure is to avoid
replication of information. The database described here
could equally well be constructed as a single file with
all the fields combined. However, this would mean that
the data which are common to a number of records, such
as all the physical properties for different batches of
the same steel, would be replicated in each record for
that steel. It is much more efficient in storage space
to put the common data in a separate file, and then to
relate these data to the other files when necessary.

To illustrate this, consider the following extract from
the database, presented as a single file and extracting
particular fields to illustrate field replication.

Steel	Batch	Density	%Cr	%Ni	U.T.S.	Ductility
EN555	1234	21.43	8.6	5.6	213.6	54.3
EN555	1235	21.43	8.54	5.76	218.4	62.1
EN555	1236	21.43	8.76	5.62	221.2	61.9
EN555	1237	21.43	8.82	5.8	214.2	60.2

It is apparent that the density field is constant in all
the records, since it depends only on the type of steel
and not the individual batch. Repeating this field means
that

a)   the information must be entered individually for every record in the database, and

b)   the database is physically larger than it need be.

It is therefore normal to place common data like this in a separate file, linked by a common field; in this case the steel type. With such a structure, some files will contain information concerned with the steel type, while other files will relate to a particular batch. When information is retrieved on a batch of steel, the details common to all batches will be retrieved from the single record in the steel-specific file.

The overall structure is shown in diagrammatic form below

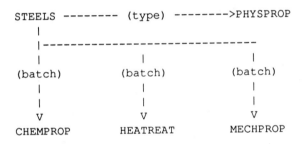

The main linking field is shown in brackets in each case.

9.2 Retrieving data from the database

Retrieving data from such a linked set of files is quite easy with modern relational database systems. Each field required is related to its file by prefixing it with the file name. For instance, using dBase (a popular database system on personal computers), a command such as

    LIST STEELS->TYPE, PHYSPROP->EXPAND, CHEMPROP->CR

will list the stated fields from three different files, i.e. the TYPE from the STEELS file, the THERMAL EXPANSION COEFFICIENT from the PHYSPROP file and the %Cr value from the CHEMPROP file, each related or linked by the fields already described.

52

Such retrieval of data is not limited to displaying or printing; it is common to select data from a database system and transfer it to a file which may then be used with a range of data analysis programmes, both general purpose, eg statistical analysis systems, and specific custom-built programmes for particular types of analysis. These facilities are a replacement for the built-in analysis functions sometimes found in larger computer systems, and can in many cases give improved flexibility.

In the example given above, the following dBase command will produce a file which can be read by a range of other programmes (assuming the current file is MECHPROP).

    COPY FIELDS UTS, MODULUS TO UTMOD.DAT DELIMITED

The file produced will be called UTMOD.DAT, and will contain the UTS and Modulus values for all data in the MECHPROP file.

9.3 Hardware and software for materials databases

A private database for materials data may be set up using almost any type of computer hardware; these applications are not particularly onerous in terms of computer size and power, and are not usually beyond the capabilities of quite modest systems. The most important criteria are the capacity of non-volatile (usually disc or tape) memory available, and the availability of suitable database software.

Given current trends in the size and power of computer systems, and their spread to the desks of users, it is more than likely that such database systems will utilise a microcomputer system. Larger systems (minis and mainframes) do have advantages in data security, but these issues which are of vital importance with huge multi-user databanks are much less so with relatively small, mostly single-user databanks such as we are concerned with here.

If a microcomputer system is used, it will most likely be an IBM PC compatible type. Such machines are

53

available with hugely varying power and speed: cheap
machines cost little more than an office typewriter, but
will still perform quite adequately, while the most
expensive machines can be as powerful as some mainframe
systems. The common database software systems will run
on all such machines, provided they have sufficient RAM
memory and disc capacity.

Concerning database software; it is impossible to make
recommendations, but following is a list of some popular
packages, and then some of the features which are likely
to be useful in maintaining a materials database.

Ashton-Tate dBase is easily the most popular general
purpose database package. It began as version II (there
was never a commercial version I) on 8-bit CP/M
machines, and was followed by version III and, recently,
version IV on 16-bit PC compatibles.

Microrim R:BASE is another very popular package, in some
ways more powerful than dBase.

Borland Paradox is a very powerful system, very good for
networking and multi-user access.

The **Smart database** is part of the most successful
integrated package for microcomputers (the other
functions are word processing, spreadsheet and
communications). It is available as a stand-alone
database and is good enough to be used as such.

The database functions of other integrated packages,
such as Lotus Symphony and Ashton-Tate Framework, are
much more limited in functionality and not suitable for
anything more than a very simple list of properties.

Also available are some clones (near copies) of dBase,
which are very similar in operation and usually much
cheaper. The best known of these are **FoxBase** and
VP Info.

The other microcomputer system which may be encountered
in a scientific environment is the Apple Macintosh. This
machine is not commonly used for large databases, its
most popular applications being in graphically

orientated uses such as presentation graphics and desktop publishing. However, several good database programmes are available, such as:

dBase Mac which is not a simple conversion of the PC product; it has a completely different method of operation.

FoxBase Mac is preferred by many experts to the Ashton Tate programme, being much faster and conforming more closely to the Macintosh standard.

Omnis is a very powerful product which can be customised to suit many applications. The normal uses of the programme probably lean more than most towards typical business uses such as accounting, ticketing etc., but this does not of course mean that the programme is unsuitable for scientific applications.

9.4 Database requirements for materials systems

Materials databanks have certain common requirements which stress particular aspects of a database system's capabilities. Some of the more important ones are listed below

a)   Flexible data retrieval. This is probably the most important feature. The data required from a materials database can vary enormously. At different times, any combination of fields may be required, as a list or a formatted report, selected on a range of different criteria. By contrast, many standard business applications have only a single output format.

b)   Powerful indexing. Materials databases usually require indexing on several fields. This is one aspect in which modern "relational" database systems score heavily over older "hierarchical" systems, since they typically allow many simultaneous indexes to be maintained concurrently.

c)   Built-in calculation capabilities. The provision of calculated fields, produced by numerical means from one or more other fields. Some systems provide rather more in the way of data analysis, such as statistical and/or regression analysis.

9.5 Data security

The security of valuable and sometimes irreplaceable data can be equally important as in the large systems mentioned earlier, although the scale of the problem is much smaller. Security in the sense of avoiding the accidental loss of data is best ensured by careful and rigorous backing up of the data, either to a tape system, to a back-up disc or to another computer, often a shared mainframe or minicomputer. In this respect a network system with a large file server is ideal, since the data can be stored on a local disc with the file server used for backup.

Facilities for ensuring the security of data in the other sense, i.e. by preventing unauthorised access, are unfortunately much more restricted in personal computer database systems than on large machines.

Some systems have a rudimentary password system, and some even use encryption of the data, but both are usually quite simple for an expert to get around. There are, however, much simpler methods of protecting data. The best is to use a removable, rather than a fixed, disc system. This may be a large capacity floppy disc, or a removable cartridge disc, or one of the new demountable Winchester disc packs.

Removing the data in this way, and keeping it securely stored away from the computer, is an excellent method of data protection.

10   COMMERCIAL MATERIALS DATABANKS

The explosive growth in available data has been one of the main trends of the 1980's, and the provision of on-line databanks, in a multitude of areas, has been one of the results of this trend. These data sources vary enormously in the scope, validity and relevance of the data they provide. Many are composed solely of manufacturer's details; others contain references to relevant literature, while still others (usually a minority) include actual independent test data.

It is always worth remembering when considering the use of a commercial databank that the provision of such data is not, except in a few cases, an act of benevolence towards a world hungry for data; these are commercial systems and are expected to make money. Some can be extremely expensive to use.

A selection of commercial databank systems is listed here, as an indication of the types of systems available. Each name has a type code after it. The types included are as follows

TYPE 1   Manufacturer's data

TYPE 2   Literature references

TYPE 3   Independent research data

10.1 General

METADEX (1)	The best-known source of bibliographic data on metals	ASM, USA
CDC (1)	Properties of copper alloys	Copper Dev. Association, USA
ZLC (3)	Mechanical properties of Zn, Pb, Cd and alloys	Zinc Dev. Ass. UK
POLYMAT (1)	Properties of polymers	DKI W Germany
MATUS	Properties of various materials; plastics, metals, ceramics, etc.	Engineering Information Company Ltd, UK
PERITUS (1)	Selection system for metals and plastics	MATSEL, UK
H DATA (3)	Hydrogen embrittlement of metals	ENSCP-CNRS France

10.2 Machinability

INFOS (3)	Cutting and machinability of metals	EXAPT-Verein W Germany
CUTDATA (3)	As above	Metcut Research Association, USA
USIDATA (3)	As above	ADEPA, France

10.3 Steel properties

BOLTS (3)	Mech props of bolting materials	EPRI, USA
HTM-DB (3)	High temperature materials, mechanical and corrosion properties. Mainly Alloy 800 at present	CEC JRC, Netherlands
METALS (2)	Compositions, mechanical and physical properties of metals	ASM, USA
STEEL FACTS/S (1)	Mechanical properties etc. of metals	BFI, W Germany
STEEL FACTS/T (2)	As STEELFACTS/S	BFI, W Germany

10.4 Access to databanks

Most of the systems listed above are available on-line, i.e. the host computer may be interrogated and the databank searched for relevant information. The mechanics of connecting to the computer vary: some are on WAN's (Wide Area Networks) which can be accessed from sites around one or more countries, while others can be called up using direct telephone lines. In either case, most systems can be accessed through PSS (Packet Switchstream Service) which appreciably reduces the connection cost.

Some systems are not on-line, but are supplied to customers on disc or tape for their own use. With the increasing popularity of personal computers, this method is becoming more common.

10.5 Networking and sharing of databanks

Several of the databanks listed above are already available, or are planned to be made available, on public or semi-public networks which are intended to provide a range of data from many computer centres. Two of the most important of these are the American NMPDN (National Material Property Data Network), set up in 1984 to integrate a range of data sources from metal producers, engineering industries and learned societies, and EURONET/DIANE, which is a European, EEC-originated, network intended for the sharing of a range of data between EEC members.

10.6 The future of databanks

In addition to an expansion of data available, which is inevitable, future trends will probably follow two main paths. These are not mutually exclusive: progress in one will largely facilitate the other.

The first is increased integration of data systems, both as regards improved accessability via networks and as regards progress towards common access methods and operating procedures (the "user interface" which is so often mentioned in connection with computer systems). This second point is one of the most important; the multiplicity of different systems and methods of accessing data have been identified time and again as a major reason for the lack of use of commercial systems.

The second route will be that towards "Expert systems", which is the true goal of many interpretive databanks, i.e. systems which incorporate selection and manipulation of raw data to produce meaningful design information. The ultimate system in this respect will be an expert system which utilises both the raw data and a knowledge of the rules for application of the raw data to real situations, and whose output will be directly

applicable to the problem in hand without further
interpretation or analysis.

An obvious use for such a system would be in the
selection of the best material for a particular
component; the physical and mechanical properties
requirements may be quite simple to select, but the
application would be complicated by the number of
properties needing to be considered, from corrosion
properties and compatibility with other adjacent
materials, through weldability and/or machinability to
such things as cost, availability and security of
supply.

Such universal expert systems are still some way off,
although a start has been made in the UK with corrosion
data.

11   STANDARDISATION OF DATA SOURCES

As well as providing facilities for linking systems and
transferring data between them, there have also been
recent moves towards standardising on data structures,
or at least on the data stored, between different data
sources. These efforts have involved the DTI in the UK,
the CEC in Europe, the National Bureau of Standards and
Boeing Corporation in the USA, and CODATA, which is a
world-wide organisation dedicated to the sharing of
advanced technology data.

This is obviously a worthwhile goal, but it is
complicated by the range of disparate computers,
operating systems and database software involved. Other
complicating factors are the differences in the
description, validity and source of data already
discussed.

12   REFERENCES

1   Codd, E F, "A relational model of data for large
     shared data banks", Comm. ACM, **13** (1970), 377-387

2   Codd, E F, "Further normalisation of the data base related model", Courant Computer Science Symposia, "Data Base Systems", New York University.

3   Codd, E F, "Relational completeness of data base sublanguages", *ibid*.

A good reference on database structures is:

Principles of database management, by James Martin, Prentice Hall 1976

Some general references to computerised materials databanks, mainly concerned with standardisation and linking of systems, are:

Factual material data banks, by H Krockel, K Reynard and G Steven, published by the Commission of the European Communities, 1984.

Materials data systems for engineering, by J H Westbrook, H Behrens, G Dathe and S Iwata, published by CODATA

Computerised materials data systems, by J H Westbrook and J R Rumble, published by the National Bureau of Standards.

3:Data analysis

L M JENKINS

Rolls Royce Plc, Bristol

SYNOPSIS

The integrity assessment of a component is a fundamental part of the design process. Materials data forms one of the major inputs into this assessment. The data includes physical and mechanical properties in addition to life estimation data. In most applications failures are at best inconvenient and usually the consequences are far worse. It is therefore necessary to design components to achieve a minimum life for a given level of risk. This requires the factoring of data from typical properties to minimum values associated with the risk level.

1. INTRODUCTION

A schematic of the design process is shown in figure 1. The presentation is the authors interpretation of the cycle and many others abound [1,2]. The procedures are applicable to most engineering products from a garden spade to a gas turbine. It is only the complexity of each stage which is different. The process starts with a specification and culminates in the product entering service with the customer. In between, the design duty leads to a proposed design which is analysed in terms of performance, integrity, cost, etc. and then followed by the product development phase to validate that the design meets the requirements of the specification. Any shortfalls identified during the analytical assessment, development or service use of the product leads to design modifications to rectify the situation.

Looking at integrity assessment in more detail (figure 2) highlights the need for materials data in a form suitable for the analysis method used. For example, a finite element model of the creep behaviour of a component requires a definition of the variation of strain with stress, time and temperature in order to calculate the rate of creep and subsequent redistribution of stresses. Data is required for each failure mechanism considered. The following is not exhaustive but identifies the major causes of component malfunction:-

1. Tensile overload leading to
   a. Unacceptable deformation
   b. Fracture

2. Fatigue - Low cycle
           - High cycle

3. Creep

4. Corrosion, Oxidation

5. Erosion

6. Interactions of mechanisms

   etc.

This paper concentrates on the analysis of fatigue and creep data.

2. STATISTICS

2.1 BACKGROUND

Statistics has been defined as 'the science of making decisions in the face of uncertainty' [3], and statistical methods are applicable whenever such uncertainties have a large effect on the phenomena being studied. This is particularly applicable to the variation of materials properties. The tensile properties of a metal are dependent on the grain size, the resistance of the individual grains to deformation and the strength of the grain boundaries. These microstructural variations are in turn dependent on other factors such as material composition and thermo-mechanical history (heat treatment, processing route etc.). It is therefore hardly surprising that supposedly identical test pieces exhibit different properties. The variation must be taken into account in order to reliably design components with suitably analysed materials data.

It is necessary to determine the shape of the distribution of properties. This can only be done accurately if there are a large number of available results. Normally, the testing of a large number of test pieces is both prohibitively expensive and time consuming. It is therefore usually necessary to obtain meaningful information about the characteristics of the population from a limited sample of the population. Statistics is based on the idea that the sample is typical in some way and that it will enable predictions to be made about the whole population. Often basic assumptions can be made about the shape of the distribution based on previous experience. Most data conforms to one of several standard probability distributions. Each of these distributions has been thoroughly investigated and their properties determined such that they have been reduced to tables for practical application

and special probability papers have been
evolved which make their interpretations
relatively simple. Having assumed the general
form of the curve, the observed results are
used to estimate the constants; mean, standard
deviation etc. which define the specific
location and spread of the distribution. It is
then possible to predict the probability of
results exceeding certain values. The
log-normal and Weibull [4] distributions have
found widespread applicability for the
analysis of materials data and are discussed
in later sections of the paper.

It is normal practice in many industries to
design components using data relevant to a so
called minimum strength part. There is in
practice the possibility of a component being
produced with a lower strength. It is the
level of acceptable risk associated with a
minimum strength part which determines the
safety factors applied to typical data. The
level of acceptable risk is dependent on the
expectations of the user and considerations of
the safety and commercial implications of
failure. The use of too conservative safety
factors results in over weight and costly
designs or a low yield of acceptable
components in production.

2.2 LOG NORMAL DISTRIBUTION

The normal distribution is probably the most
important and certainly the most commonly used
theoretical distribution, its physical
appearance is a symmetrical bell shaped curve,
tailing to infinity in both the positive and
negative directions. Many of the distributions
that occur in practice are roughly normal,
having a single centrally situated hump,
approximate symmetry and long tails. Often a
pattern of observations is obtained which
cannot be reasonably attributed to the normal
distribution. In this case, it is of great
advantage if , by a simple transformation of
the variable, the characteristics of the

normal can be used. One commonly occurring distribution that can be readily transformed is the log normal distribution. This typifies many fatigue failure situations where the distribution is found to be normal if the log of the fatigue life is plotted.

The standard deviation is a measure of the scatter in the test results and their deviation from the mean. One characteristic of a normal distribution is the area below the curve representing a proportion of the total population bounded between the mean and n standard deviations is independent of the spread or range of the distribution. For example, 68.26% of a normal distribution will be within ±1 standard deviations of the mean, 99.44% within ±2 standard deviations and 99.73% within ±3 standard deviations (figure 3). It therefore follows that 739 in 740 results lie above the -3 standard deviations point in the distribution. For many aerospace applications the ratio across 6 standard deviations (±3) is taken as being effectively the full width of the distribution, viz

$$\log N_{max} - \log N_{min} = 6 * \log N_{std\ dev}$$

transforms to $\quad N_{max} / N_{min} = N_{std\ dev}^{\ 6}$

and hence $\quad N_{std\ dev} = (N_{max} / N_{min})^{1/6}$

Similarly the difference between N_{mean} and N_{min} yields

$$\log N_{mean} - \log N_{min} = 3 * \log N_{std\ dev}$$

therefore

$$\log N_{mean} - \log N_{min} = 3 * \log(N_{max}/N_{min})^{1/6}$$

and finally $\quad N_{mean} / N_{min} = \sqrt{N_{max} / N_{min}}$

66

For a ratio of Nmax to Nmin of 6 the ratio of Nmean to Nmin is $\sqrt{6}$ (2.45).

It should be noted that the 1:740 statistics in the aerospace industry applies to the production of a small crack or reaching a proportion of the failure life and not the life to dysfunction. The overall risk of failure is therefore greatly less than this figure as testified by service experience.

2.3 WEIBULL DISTRIBUTION

The Weibull probability curve is commonly used to describe the distribution of times to failure in fatigue and creep testing and also the distributions of service lives of many components. Each of these represent how failures are distributed over some time scale such as cycles, hours etc. This type of analysis is amenable to predicting the number of failures which might be expected after a certain time or number of cycles. Alternatively the variable does not need to be time based but can be used to handle other parameters such as component dimensions or crack sizes as discussed in section 3.4.

The three parameter distribution proposed by Weibull has great flexibility because altering the three parameters enables it to take many different shapes and move its central location. The distribution is described mathematically [4] by:-

$$F(t) = 1 - e^{-((t-t_o)/\eta)^\beta}$$

where $F(t)$ = fraction failing
t = failure time
t_o = origin of the distribution
η = characteristic life
β = slope or shape parameter
e = exponential

t_o, η, and β are the three variable parameters.

67

It is often convenient to initially perform a two parameter Weibull analysis (i.e $t_o = 0$) to estimate a reasonable initial estimate of t_o from the lower end asymptotic value of the distribution before attempting a three parameter analysis.

Because of the complexity of the above equation special plotting paper has been produced, essentially log log versus log scales, which permits complex distributions to be plotted as straight lines (see figure 4). The first task in a Weibull analysis is to arrange the test results in ascending order. For a population containing only failed results the rank order number assigned to each result is simply the order of the ascending test results. It is then necessary to assign a probability to each measurement; that is to estimate the value on the probability scale against which each measurement will be plotted. Statisticians have produced tables of values called Median Ranks which are the probabilities that are most likely to occur when a particular number of results are plotted. The Median Ranks vary with population size. For large populations the Median Ranks can be obtained with acceptable accuracy using Bernard's formula, viz

$$M = r - 0.3 / (n + 0.4)$$

where   M = Median Rank
             r = rank order number
             n = number of results in total
                 population

A plot of Median ranks against $(t - t_o)$ is produced for a full three parameter Weibull analysis. Non failed results can also be included in the analysis as suspended points by assigning non integer rank order increments to the failed test results. The methodology is beyond the scope of this paper and should be referred to elsewhere [4].

3. FATIGUE

3.1 LOW CYCLE FATIGUE

The usual representation of low cycle fatigue data is shown in the S-N curves of figure 5. Life prediction is essentially a relatively simple process of calculating a peak stress and reading off the life from a curve for an appropriate stress gradient. The curves can be based on laboratory specimens or full scale component tests or a mixture but any predictions based solely on specimens are potentially unsafe [5]. The necessity for component testing is for two main reasons. Firstly the surface condition of the test vehicle is important. Components and ideally the laboratory specimens should have representative surface finishes. Additionally, residual stresses from the manufacturing process need to be quantified to permit interpretation of the test results. Secondly, the volume/surface area subjected to high stress levels significantly affects the number of potential initiation sites which can be exploited. Statistical techniques have been used to explain the discrepancies between specimens and rotating engine disc rig results where the ratio of highly stressed material was about 1:50 [6].

The results of a laboratory specimen fatigue test programme are reproduced in figure 6. The scatter in the data is clearly visible. It is necessary to fit a line to the mean of the data firstly to permit interpolation between test conditions and secondly to adjust to minimum properties using suitable factors. One fit often used is a power law of the form:-

$$C = N.\sigma^n$$

This is achieved by plotting log stress against log life (figure 7) and attempting a fit to the data using linear regression

69

analysis. The low stress long life results are indicated as unfailed test pieces. The exponent n is equal to minus the inverse of the gradient and the constant C equal to the intercept of the line with the one cycle life line. The approach tends to produce a curve which is not a good fit to the data and does not reproduce the run out seen for many materials at lower stresses which is due to the increase in crack initiation life as a proportion of total life with decreasing stress level. As a consequence the power law is often modified to the form:-

$$C = N.(\sigma - \sigma')^n$$

where σ' is a pseudo run out stress but is generally chosen to give the best fit over the range of available data. The fitting is achieved by plotting $\log (\sigma-\sigma')$ against $\log N$ and varying σ' to optimise the fit in the upper plot of figure 8. The data plotted in figure 6 has been replotted in the lower plot of figure 8 for different values of σ'.

Generally test results are a mixture of failed and unbroken results. The unbroken test piece results are of limited use but do give an indication of the shape of the S-N curve especially in the run out region.

To produce minimum properties data it is necessary to factor down the mean line and ensure that it envelopes all of the scatter present in the results. For many gas turbine materials a factor of 2.45 is applied to the life and a factor of 1.18 applied to the stresses. This is based on a maximum to minimum population variation of 6 on life and 1.4 on strength assuming a log-normal distribution. The minimum properties line is the envelope of the lower bound of factoring on life or stress (figure 9).

3.2 CRACK PROPAGATION

A number of alternative types of crack propagation specimen types are available (eg. compact tension, corner crack, etc.). In each case the progress of the crack is recorded using electrical potential drop monitoring techniques [7,8] to produce a tabulation of crack depth versus number of test cycles. The slope of the a v. N curve is determined to produce the crack growth rate, da/dN, at each crack depth measurement. One way this is achieved is by fitting a straight line to a running number of sequential test points using a least squares fit and calculating the gradient. If the regression coefficient of the fit is below an acceptable value the extremes of the sample are eliminated until sufficient improvement is achieved. For the next point the sample is then upgraded by including the next test reading and dropping the first point of the previous sample. Typically seventeen readings might be considered to smooth out scatter due to experimental processes. For standard test piece geometries calibration curves of normalised stress intensity factor versus normalised crack depth exist based on various analytical solution techniques including finite elements [9]. Figure 10 shows the so called compliance function for a compact tension specimen [10]. Using the compliance function it is possible to calculate the stress intensity factor range ΔK for a given crack depth and hence plot da/dN versus ΔK for the duration of the test which is the usual method for presenting such data (figure 11).

Various types of fit to the crack propagation test data are used in life calculation programmes. The following relationships are some of those used:-

Paris [11] $da/dN = C.\Delta K^n$

71

where ΔK = stress intensity range of the loading cycle

$= K_{max} - K_{min}$

Forman [12] $da/dN = C.\Delta K^n / ((1-R).K_c - \Delta K)$

and R = stress intensity ratio K_{min} / K_{max}

K_c = material fracture toughness

A modification proposed by McEvily has been used where

$$da/dN =$$

$$C.(\Delta K - \Delta K_{th})^n / ((1-R).K_c - \Delta K)$$

and ΔK_{th} = threshold stress intensity range below which crack growth is negligible

Coles [13] $da/dN =$

$$e^B.(\Delta K/\Delta K_{th})^P (\ln(\Delta K/\Delta K_{th}))^Q (\ln(K_c/\Delta K))^D$$

Wallace [14] $da/dN =$

$$10^{(C_1 \sinh(C_2(\log\Delta K+C_3))+C_4)}$$

Tabulation da/dN, ΔK pairs

C, n, B, P, Q, D, C_1, C_2, C_3, C_4 are all constants applicable to the particular equation used to fit the data.

3.3 FRACTURE MECHANICS LIFE PREDICTION

An example of the structure of a fracture mechanics life prediction program is shown overleaf.

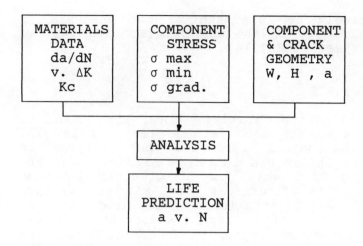

where W and H are characteristic dimensions of the component and a is the crack dimension (depth/length).

The analysis phase is further expanded in the following schematic.

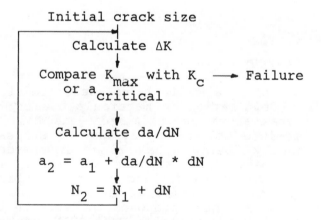

For a specified initial crack size the life to failure can be calculated. It therefore follows that if the life to failure or end of test condition is known it is possible to back calculate to the initial crack size. A relatively new approach to the analysis of fatigue data is the use of effective initial flaw sizes (EIFS)[15].

3.4 E.I.F.S

The effective initial flaw size can be calculated for each fatigue test result. The distribution of EIFS can be analysed to predict the mean and maximum EIFS's. It is the maximum EIFS which is of greatest interest since this determines the minimum fatigue life. Weibull statistics are used to plot cumulative probability against the inverse EIFS as shown typically in figure 12. The lower asymptote tends to the maximum EIFS. The maximum EIFS is then used in a fracture mechanics calculation to produce lifing design curves which can have further safety factors applied if required. For materials which exhibit significant initiation life variation with stress this needs to be built in to calculations separately.

4. CREEP

4.1 BACKGROUND

Creep is the term used to describe time dependent deformation in materials usually at temperatures above approximately one half of their absolute melting point and at stresses significantly below the time independent tensile capabilities of the material. The phenomenon of creep is due to the combined effects of time at temperature and stress. A number of models have proposed to describe the relationship between strain, time, stress and temperature.

The classical view is of creep occurring in a number of discrete phases; an instantaneous extension (elastic plus inelastic strain); a transient (or primary) creep of decreasing rate; a steady state (or secondary) creep, approximately linear with time; and an accelerating (tertiary) stage leading up to final fracture (figure 13)[16]. In practice this is a gross oversimplification and some materials do not exhibit certain phases of the above description. For example austenitic steels exhibit significant secondary creep as the effects of recovery and work hardening on strain rate negate each other while in nickel superalloys the secondary phase is almost non existent.

There are two different approaches to the analysis of creep data. The first involves an understanding of the metallurgical fundamentals at the microscopic level from which the material behaviour is predicted. The second is empirical and relies on the observed macroscopic phenomenon being described in mathematical terms. It is the empirical approach that is of most use to the engineer or designer in a commercially orientated environment. However, the understanding of the mechanism of creep is also vitally important if progress is to be made in the characterisation of material behaviour.

Creep is basically a thermally activated process such that thermal energy is required to overcome obstacles to plastic deformation such as dislocation intersections and strengthening precipitates. The strain rate is often represented by an equation of the form:-

$$\dot{\varepsilon} = A.\exp^{-q/kT}$$

where   A is a material constant
          q is the activation energy required
            to overcome the obstacle
          k is Boltzmann's constant
          T is the absolute temperature

It is well documented that crystalline metals deform by the following types of mechanism:-

1. Dislocation glide - the movement of dislocations by glide along slip planes

2. Dislocations overcoming obstacles by
   a. shearing through a precipitate
   b. climb over the obstruction by diffusion of vacancies
   c. bowing of the dislocation when the separation of the obstacle is sufficiently large

3. Shear on, or adjacent to grain boundaries

4. Vacancy diffusion inside the grains

The activation of the different creep mechanisms are not usually mutually exclusive but tend to be present in varying degrees with a particular mode of plastic deformation dominant. This concept has led to the production of deformation mechanism maps (figure 14)[17]. A map is a diagram with stress and temperature axes divided into fields within which a particular mechanism is dominant. For example shearing of precipitates tends to occur at high stresses while climb occurs at higher temperatures and lower stresses. At even higher temperatures the bowing of dislocations between precipitates is made possible as the precipitates coarsen thus increasing their spacing or lose their coherence with the surrounding matrix. Also superimposed on the maps are contours of constant strain rate. The maps are constructed from rate equations, one for each mechanism. The rate equations are functions of strain rate, stress, temperature and internal structure parameters such as grain size, precipitate size and spacing, dislocation density etc. The aim is to predict the deformation mechanism maps from fundamental principles. However, the measurement of the different variables is complex, sometimes impossible and results in adjustments to parameters in order to obtain a good fit to the experimental data. It is these problems which lead to the use of empirical models as discussed later in section 4.2.

Figure 15 shows an idealised stress rupture diagram indicating the slopes of the curves as determined by the Graham and Walles approach [18]. In the different regimes the following types of behaviour are observed:-

n = 4 Grain boundary cavitation
n = 8 Triple point cracking

Both produce many small cracks which coalesce to cause failure.

n = 16 Intergranular cracking along grain
          boundaries generally on the surface
          leading to single large crack.

In fatigue and low temperature creep transgranular cracking is more usually observed.

In reality the stress rupture diagram shows a much smoother transition between mechanisms as they interact with the resultant behaviour being a summation of the different rate equations for each mechanism.

4.2 EMPIRICAL CREEP MODELLING

As described earlier the prediction of a materials creep behaviour using theoretical techniques is fraught with practical difficulties, not least the effects of changing alloy content, material processing or time dependent microstructural changes during creep.

Empirical methods rely on the use of mathematical relationships which map the variation of strain with stress, temperature and time. The following list is a sample of the models extracted from the literature.

The θ (theta) projection method was proposed by Evans and Wilshere [19] and is of the form:-

$$\varepsilon = \theta_1(1 - e^{-\theta_2 t}) + \theta_3(e^{\theta_4 t} - 1)$$

The equation describes the strain as a sum of a decaying primary (first term) and accelerating tertiary (second term) component. θ_1 and θ_3 act as scaling terms defining the extent of the primary and tertiary stages with respect to strain, while θ_2 and θ_4 are rate parameters governing the curvatures of the primary and tertiary stages respectively.

A fit is obtained to a matrix of experimental test results covering both the stress and temperature domain such that

$$\log \theta_i = a_i + b_i \sigma + c_i T + d_i \sigma T$$

for i = 1 to 4

Thus a total of sixteen constants are required to predict the full creep map. These can be obtained using numerical methods coded into a computer algorithm to optimise the fit to the experimental data. A least squares regression analysis is often performed or alternatively an iterative process followed from an initial estimate of some of the constants. It should be noted that in some curve fitting implementations the assumption is made that b_i equals a constant * d_i and thus only twelve independent variables need to be calculated.

Other models include

$$\varepsilon = At^{1/3} + Bt + Ct^3 \qquad [20]$$

which is a special case of the Graham and Walles equation [18]

$$\varepsilon = \Sigma \ C_i \ \sigma^{\beta_i} \ (t \ (T' - T)^{-20})^{k_i}$$

$$\text{for } i = 1 \text{ to } r$$

where $k_i = ...1/3, 1, 3....$

$$\beta_i / k_i = 1, 2, 4, 8, 16....$$

Hyperbolic creep law

$$\varepsilon = A \ \ln^2(t/P) + b \ \ln^2(E/Q)$$

$$+ C \ \ln(t/P) \ \ln(E/Q) = R^2$$

Hyperfuction creep law

$$\varepsilon = A \ \tanh \ (Bt) + C \ \sinh \ (Dt)$$

Cubic creep law

$$\varepsilon = A \ + Bt + Ct^3$$

and in a more general form

$$\varepsilon = At^m + Bt^n$$

where $m<1, n>1$

The calculation of the constants in the various equations is similar to that described for the theta projection method.

The stresses in a component and their relaxation due to creep can be calculated using one of the commercially available non linear finite element packages. Creep data is generally input as a separate subroutine to the program. The strain in the component is calculated incrementally such that

$$\varepsilon_n = \Sigma \ (\ \dot{\varepsilon}_k \ \Delta t_k \ )$$

$$\text{for } n = 1 \text{ to } k$$

where $\dot{\varepsilon}_k$ is the strain rate at the start of time step Δt_k.

79

It is necessary to ensure that Δt is
sufficiently small to ensure a convergent
solution but in many programs the time step is
set automatically.

In a component experiencing a varying stress
and temperature history it is necessary to
employ either a time hardening or strain
hardening transition between stress and
temperature states (figure 13). Further the
interpolation of data within the bounds of the
test matrix used to produce the creep law is
relatively straight forward while
extrapolation can lead to numerical problems
in addition to being wildly inaccurate.

In some applications the use of a full blown
finite element analysis may not be necessary
or justified. It may be sufficient to
calculate a mean stress acting in the
component and compare the time to produce a
finite strain with a service requirement. For
example in power generating plant a steam pipe
is designed to last 100,000 hours with an
allowable plastic strain not to exceed one per
cent [21].

5. LINEAR DAMAGE MODELS

The concept of linear damage in fatigue was
reported by Miner [22] in 1945. His hypothesis
was based on the observed behaviour of an
aluminium alloy using flat sheet specimens
subjected to cycles of different stress ratio.
He is often credited with the relationship:-

$$\Sigma\ n\ /\ N = 1$$

where   n = number of applied load cycles
          N = cyclic life to failure at that
              load

i.e $n_1\ /\ N_1 + n_2\ /\ N_2 + \ldots + n_n\ /\ N_n = 1$

The assumption is that the damage accrued is a

linear summation of the damage due to different stress range and ratio cycles. In many cases retardation and acceleration effects are observed due to over and underload conditions. In spite of these drawbacks the use of the relationship is widespread.

The use of linear damage rules was further developed to include creep and is often referred to as Robinson's law:-

$$\Sigma \, n \, / \, N + \Sigma \, t \, / \, t_f = 1$$

where t = time on load
              t_f = time to failure at that load

The creep part of this latter equation can be modified using the Larson Miller parameter:-

$$P = T(20 + \log t_f)10^{-3}$$

where P = Larson Miller parameter
            T = absolute temperature (K)
            t_f = time to failure (hrs)

Alternative values to 20 have been used in some analyses.

The relationship of the Larson Miller parameter has been observed to be a linear function of stress although for some materials a higher order polynomial is required. If the relationship below is assumed:-

$$\sigma = aP + b$$

where a and b are constants then the Larson Miller equation can be rearranged to yield:-

$$t_f = \exp\left(\left(\left(\sigma - b\right) / \left(a * T * 10^{-3}\right)\right)\right) - 20)$$

For a given combination of stress and temperature it is a simple matter to calculate the life to failure at that condition. Numerical integration of $\Sigma \, t \, / \, t_f$ is therefore possible through a varying stress and temperature history.

The research and development of non linear
models is and continues to be a major area of
investigation.

6. SUMMARY

The author has attempted in this paper to
bridge the gap between the generation of
materials data in a laboratory environment and
its ultimate use in the design process to
clear components into service. The third
seminar "Mechanical Testing" of the
Characterisation of High Temperature Materials
is particularly pertinent to this subject and
the proceedings provide a valuable source of
complementary information. The end user of
data should be aware of many of the subtleties
and complications of materials testing and the
subsequent conversion of the raw data into a
usable form.

7. ACKNOWLEDGEMENTS

The author wishes to thank Rolls Royce plc for
permission to publish this paper and his
colleagues for the provision of supporting
information. He is further indebted to
Dr.G.F.Harrison for his advice on various
aspects of this paper.

8. REFERENCES

1.   P.Cooley;   "Engineering   Drawing,
Communication and Design", 1972, Pitman
Publishing.

2. Fatigue Design Handbook, 2nd edition,
AE-10P, SAE, 1988, ISBN 0 89883 011 7.

3. C.Chatfield; Statistics for technology, 2nd
edition, 1978, Chapman and Hall. ISBN
0412157500.

4. R.Abernethy et al.; Weibull Analysis Handbook, 1983, AFWAL-TR-2079

5. L.M.Jenkins; "Design and life assessment of critical aero-engine components", The role of advanced design methods in the life substantiation of aerospace components, 1988, I.Mech.E.

6. G.Asquith and A.C.Pickard; "Fatigue testing of gas turbine components" in "Full scale testing of components and structures",ed. K.J.Marsh, Butterworths, to be published.

7. R.P.Skelton; Fatigue crack growth (section 4.3), Mechanical Testing, Characterisation of high temperature materials, 1988, IoM.

8. M.A.Hicks and A.C.Pickard, "A comparison of theoretical and experimental methods of calibrating the electrical potential drop technique for crack length determination", International Journal of Fracture, 1982, Vol. 20 p91.

9. A.C.Pickard; "The application of three dimensional finite element methods to fracture mechanics and fatigue life prediction", 1986, EMAS.

10. E.F.Walker and M.J.May; "Compliance functions for various types of test specimen geometry", BISRA open report MG/E/307/67, 1967.

11. P.C.Paris; "The fracture mechanics approach to fatigue", 1963.

12. G.Forman et al.; "Numerical analysis of crack propagation in cyclically loaded structures", Trans. ASME D, 89, 1967.

13. Shahani and Popp, "Evaluation of cyclic behaviour of aircraft turbine disk alloys",NASA-CP-159433.

14. Cowles, Sims, Warren and Miner, "Cyclic behaviour of turbine disc alloys at 650°C", ASME 356, Journal of engineering materials and technology, Vol. 102, 1982.

15. A.C.Pickard, "Design and life assessment of aero engine components" in "Recent advances in design procedures for high temperature plant", I.Mech.E symposium, pub. MEP, 1988, ISBN 0852986807.

16. R.Reed-Hill, Physical metallurgy principles, pub. D Van Nostrand Co, Inc.,1967.

17. H.J.Frost and M.F.Ashby, "Deformation mechanism maps for pure iron, two austenitic stainless steels and a low alloy ferritic steel", Cambridge University.

18. K.F.A.Walles, "Random and systematic factors in the scatter of creep data", NGTE report No.R280, 1966.

19. R.W.Evans and B.Wilshere, "Creep of metals and alloys", IoM, London, 1985.

20. S.G.R.Brown, "Extrapolation by creep curve shape analysis", Proceedings of 3rd International conference on creep and fracture of engineering materials and structures, IoM, 1987, ISBN 0 904357 996.

21. J.D.Parker, "Life assessment in power plant components operating in the creep range", Proceedings of 3rd International conference on creep and fracture of engineering materials and structures, IoM, 1987, ISBN 0 904357 996.

22. M.A.Miner, "Cumulative damage in fatigue", Journal of applied mechanics, Vol. 12, 1945.

THE DESIGN PROCESS

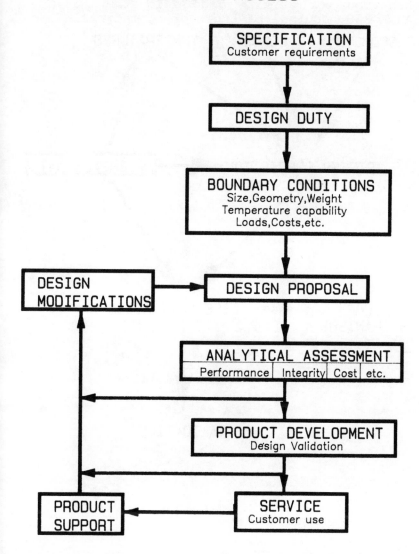

FIGURE 1

INTEGRITY ASSESSMENT

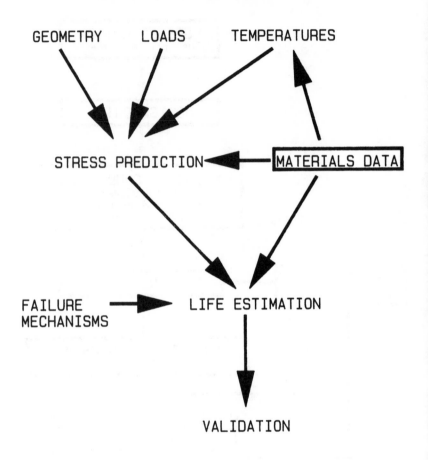

FIGURE 2.

LOG NORMAL DISTRIBUTION

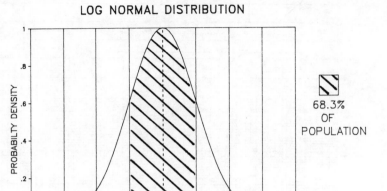

68.3%
OF
POPULATION

LOG NORMAL DISTRIBUTION

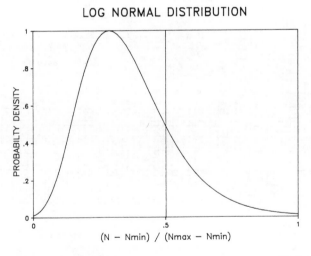

FIGURE 3

87

WEIBULL ANALYSIS

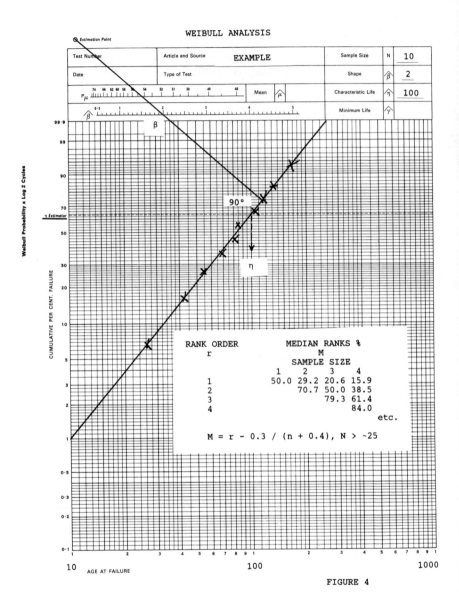

FIGURE 4

88

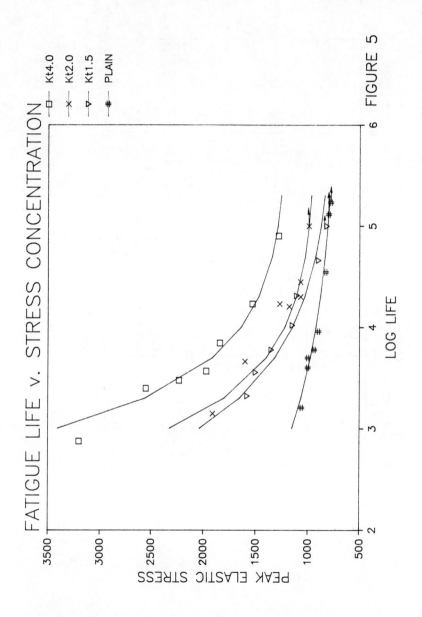

FATIGUE LIFE v. STRESS CONCENTRATION

Kt4.0
Kt2.0
Kt1.5
PLAIN

PEAK ELASTIC STRESS

LOG LIFE

FIGURE 5

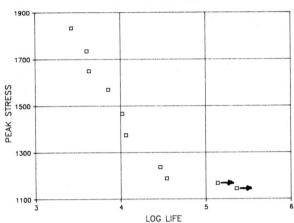

FIGURE 6

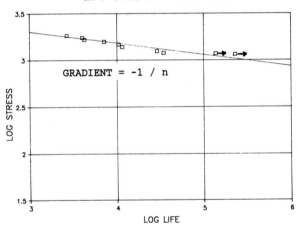

FIGURE 7

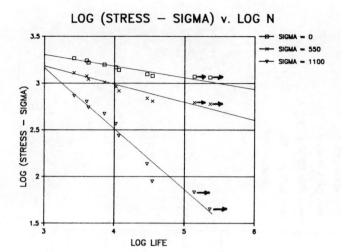

LOG (STRESS – SIGMA) v. LOG N

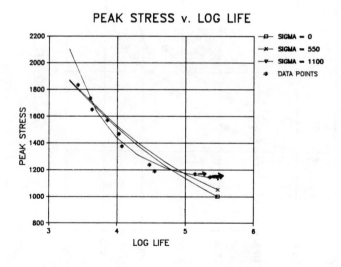

PEAK STRESS v. LOG LIFE

FIGURE 8

91

MIMIMUM FATIGUE STRENGTH

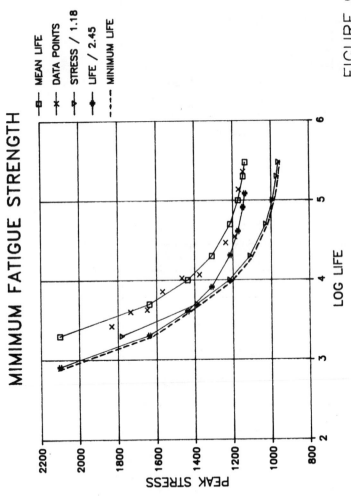

- MEAN LIFE
- DATA POINTS
- STRESS / 1.18
- LIFE / 2.45
- MINIMUM LIFE

LOG LIFE

PEAK STRESS

FIGURE 9

COMPLIANCE FUNCTION FOR
THE COMPACT TENSION SPECIMEN

$$Y = \frac{B W . K}{P . \sqrt{\pi a}}$$

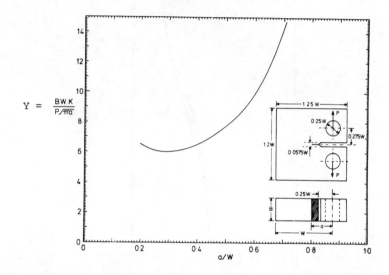

FIGURE 10

CRACK PROPAGATION DATA

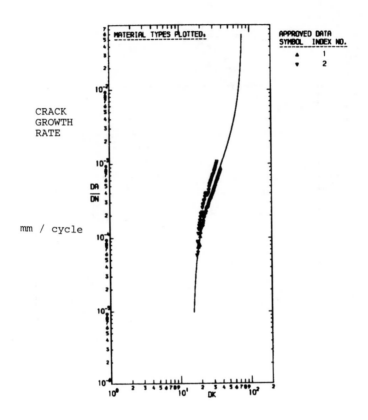

STRESS INTENSITY RANGE

FIGURE 11

94

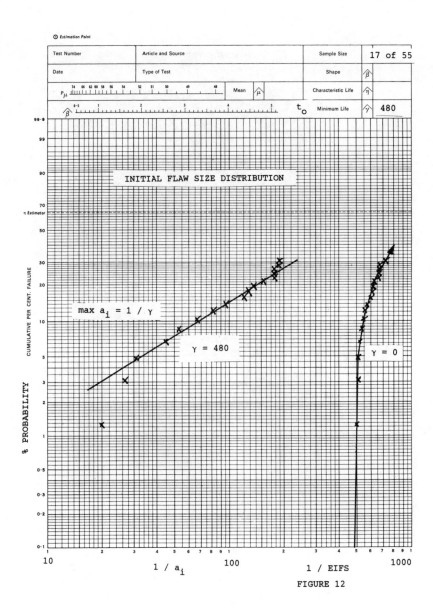

FIGURE 12

IDEALISED CREEP CURVES

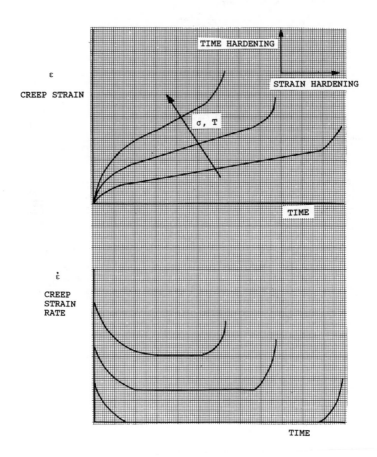

FIGURE 13

96

DEFORMATION MECHANISM MAP

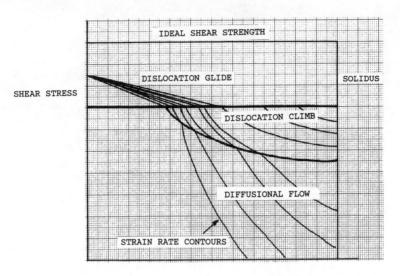

IDEAL SHEAR STRENGTH

DISLOCATION GLIDE

SOLIDUS

SHEAR STRESS

DISLOCATION CLIMB

DIFFUSIONAL FLOW

STRAIN RATE CONTOURS

TEMPERATURE

FIGURE 14

STRESS RUPTURE DIAGRAM

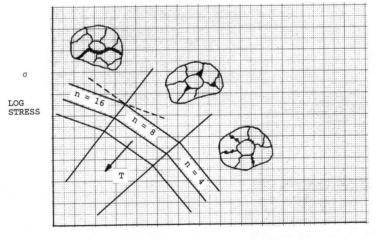

σ

LOG
STRESS

n = 16

n = 8

n = 4

T

LOG tf

FIGURE 15

97

SESSION 2
Chairman: P Spilling
(Rolls Royce, Derby)

4:Introduction to
numerical analysis techniques

R W EVANS

University College of Swansea

1. INTRODUCTION

Many situations in both theoretical and process materials technology
involve the solution of field problems. The fields which exist in
bodies can be of many kinds. Thus during commercial heat treatment of metals, knowledge of the precise temperature field inside the
workpiece is of vital importance, and the same is true of materials
during casting. In this latter case the flow field is also of interest. In
many mechanical applications stresses and strains are important field
variables and in addition to all these physical quantities the detailed
microscopic structure may be critical. Until 20 years ago many of
the complex calculations concerned with understanding materials
used an analytical approach. Quantities such as the stress field
around a dislocation, the diffusion environment around a precipitate
or the thermal gradient through a quenched metal were calculated by
solving the underlying partial differential equations using known
solutions in terms of functions or power series. The major disadvantage of these methods was that they could only deal with grossly
simplified problems, particularly with respect to geometry and
material properties. As a result of the rapid advance in computing
power over recent times, many of these problems are now tackled by
numerical procedures which are not restricted by the difficulties
encountered in analysis. However, if the techniques are to be
properly applied to yield useful results it is important that there is
good understanding of the methods, and this paper is intended to give
an introduction to such numerical techniques. The author agrees
with Lord Kelvin's view, "I have no satisfaction in formulae unless I
feel their numerical magnitude". Hence much of the analysis is
carried out in terms of numerical examples which the reader is urged
to follow through.

2. AN EARLY EXAMPLE

It is interesting to note that the numerical analysis of mathematical and engineering problems is not a new phenomenon. The procedures predate analytical methods (and in particular the calculus) by several thousand years. An early example is the solution of a problem which faced many early engineers and builders, namely the relationship between the circumference and diameter of a circle. The first reasonable values of this quantity (the value π) were derived by Archimedes using a typical numerical method.

Archimedes noted that if a regular polygon were inscribed in a circle, then the sum of the lengths of the edges of the polygon (l_i) was always less than the circumference of the circle. Figure 1 illustrates this for polygons of four and eight sides. On the other hand, the sum of the lengths of the sides of an exscribed polygon (l_o) always exceeded the circumference of the circle. Thus, if d were the diameter of the circle,

$$l_i < \pi d < l_o \ . \tag{1}$$

The value of π could thus be bounded below and above if the circumference of the polygons could be estimated. This latter task was straightforward since the sides of triangles could be accurately calculated from applications of Pythagoras' theorem. Furthermore it seems intuitively reasonable that as the number of sides in the regular polygons increases, then the width of the bound on π should decrease. Table 1 shows a series of numerical calculations carried out to 3 decimal places, and indicates the degree of approximation. The process appears to be *converging* to the value of π as n increases.

This early numerical procedure contains many of the important elements of the complex numerical methods used today. They may be summarized as follows.

1. The complex intractable problem (calculation of a curved length) is replaced by problems which are tractable (calculation of the lengths of straight lines).

2. In order to make this process reasonably accurate, the linear approximations are only applied over limited regions of the problem. Thus the overall estimation is replaced by a whole series of simple approximations. The problem is said to have been *discretized*.

3. The discretization procedure has led to an answer that is not precise but which always contains an error.

4. It is the job of the numerical analyst to provide bounds for the error and to utilize procedures which will minimise it. Such procedures are nearly always available. In this simple case they merely involved an increase in n.

5. In all numerical procedures, sufficient accuracy is only obtained if very many calculations are carried out (i.e. discretization is on a fine scale).

It is, of course, this latter problem which led to the demise of numerical methods and the popularity of analysis and the solution of complex equations. Although in these cases the analysis might be difficult, the numerical labour was small. The advent of large scale, cheap computing facilities has altered the environment for numerical work, and many of the traditional materials problems are now being solved by simple repetitive calculation.

3. GENERAL PROCEDURES

It is convenient to divide the field problems encountered in materials engineering into various types, and a classification into boundary and initial value problems is useful. They may be described as follows.

1) **Boundary value problems:** Conditions at the boundary of the material are known and it is required to calculate the values of the field variable at all interior points. Thus if a metal plate has the temperatures at its edge fixed, then the internal points in the plate will take up equilibrium values of temperature and these are the field variables to be calculated. In a similar way, if the edges are loaded with external forces, then the displacements at each point within the body represent the field variables to be calculated. Once these are known, other derivable quantities such as stress and strain can be calculated. It is clear that the boundary value problems are effectively equilibrium problems and the element of time dependence does not enter.

2) **Initial value problems:** Values of the field variable are known at all points in the body at some starting point. Then, subject to some change in boundary conditions, the field variable will change with time (or some time-like parameter) and the initial value problem is to calculate these changes in such variables. For the example above of a metal plate, one might imagine a sudden change in boundary temperatures. The internal temperatures will then change with time and the calculation of these changes constitutes the initial value problem. In the mechanical case, displacements may change with time due to creep of the solid. In many cases, the initial value problem represents an approach to a new equilibrium boundary value

problem but this is not always so. For instance, if a forging process is to be modelled, no equilibrium displacements are obtained but the variation of displacements in the body with punch stroke represents an initial value problem. For those problems which represent transient behaviour towards new equilibria, solutions to the new boundary value problem will be the same as those for the initial value problem for very long times.

The first step in the solution procedure for each of these problem types is the discretization of the space. This is usually done by dividing up the continuum into small volumes as shown in Figure 2. The process can be carried out in a number of possible ways. The shapes of the small volumes can be quite general, and Figures 2a and b show examples for two dimensions in the form of squares or triangles respectively. In some cases the field variable is assumed to be constant over the small volume and in this case the value of the field variable is assigned to a point inside the volume. Thus, in Figure 2a each point might be assigned its own temperature. In other cases, the field variable is assumed to vary as a simple polynomial across the volume (e.g. linearly) and in this case the values of the variable must be assigned to points on the edge of the discretized volume. Thus, in Figure 2b, the triangles may have a linear variation in temperature in which case the values of temperature at the corner points are sufficient to define the variation across the volume. In either of these cases the points are called *nodes*.

The field problem is now reduced to determining the values of the field variable at the nodes only. It is important to realize that this is a result of discretization. The problem has been reduced from one of finding, say, temperatures at all points to the temperatures at a fixed (possibly large) number of points. In all cases, the numerical procedure now reduces the problem to that of solving a limited number of simple equations in the limited number of unknown field variables at the nodal points. Whichever detailed method is used the problem boils down to a solution of the equation

$$\mathbf{K}\,\mathbf{a} = \mathbf{f} \qquad\qquad (2)$$

Here, $\mathbf{a}$ represents a column matrix which is just a list of the unknowns (for instance, temperature) for all nodal points. The column matrix $\mathbf{f}$ is a set of numbers which are derived from the boundary and initial conditions. $\mathbf{K}$ is a square matrix which is an array of coefficients in the equation set relating to the unknowns $\mathbf{a}$. It is often called the stiffness matrix for the problem, but this does not imply a mechanical problem. The ways in which $\mathbf{K}$ and $\mathbf{f}$ are derived depends on the solution method type and the problem type, and these will be illustrated in detail below.

In many linear cases, **K** is just a set of constants and the equations are a set of linear equations. Examples are linear elasticity and constant conductivity heat flow. In other cases (e.g. creep and plasticity) **K** may be a function of **a** so a non-linear equation set is obtained. In either case an efficient equation solver must be available since there may be many thousands of variables in **a**. Most equation solvers are based on Gauss elimination. The method uses the fact that in linear equation sets, equations can be multiplied through by constants or added to each other without altering the values of the roots. Elimination uses these properties to eliminate variables between equations. The method is best illustrated numerically. Consider the equation set

$$5a_1 + 1a_2 + 0a_3 = 7 \quad (1)$$
$$3a_1 + 6a_2 + 1a_3 = 18 \quad (2)$$
$$1a_1 + 2a_2 + 4a_3 = 17 \quad (3)$$

i.e.

$$\begin{bmatrix} 5 & 1 & 0 \\ 3 & 6 & 1 \\ 1 & 2 & 4 \end{bmatrix} \begin{bmatrix} a_1 \\ a_2 \\ a_3 \end{bmatrix} = \begin{bmatrix} 7 \\ 18 \\ 17 \end{bmatrix} \quad (3)$$

Now form a multiplier, m, by dividing K_{12} by K_{11} (i.e. $m = 3/5$). Multiply equation *(1)* by m and subtract it from equation *(2)*. In a similar way, redefine $m = K_{31}/K_{11} = 1/5$, multiply equation *(1)* by m and subtract from equation *(3)*. The equation set now takes the form

$$
\begin{array}{llll}
5\,a_1 & + 1\,a_2 & + 0\,a_3 & = 7 & \qquad (1) \\
\hline
0\,a_1 & + 27/5\,a_2 & + 1\,a_3 & = 69/5 & \qquad (2) \\
0\,a_1 & + 9/5\,a_2 & + 4\,a_3 & = 68/5 & \qquad (3)
\end{array}
\quad (4)
$$

This completes the first elimination block, and a_1 has been eliminated from equations *(2)* and *(3)*. The process can now be repeated on the sub-matrix of equations below the line in equation (4). Thus, set $m = K_{32}/K_{22} = 1/3$, multiply equation *(2)* by m and subtract from *(3)*. The equations are now

$$
\begin{array}{llll}
5\,a_1 & + 1\,a_2 & + 0\,a_3 & = 7 & \qquad (1) \\
0\,a_1 & + 27/5\,a_2 & + 1\,a_3 & = 69/5 & \qquad (2) \\
0\,a_1 & + 0\,a_2 & + 11/3\,a_3 & = 11 & \qquad (3)
\end{array}
\quad (5)
$$

Note that equation *(3)* now only contains a_3 so that a_3 can be obtained directly. This value can then be backsubstituted into equation *(2)* to obtain a_2 and so on. When this is done,

$$a_3 = 3, \quad a_2 = 5/27\,([69/5]-3) = 2 \quad a_1 = 1/5\,(7-2) = 1 \quad (6)$$

The important point to note is that the solution process is highly repetitive and hence ideal for computer implementation. The following simple FORTRAN program will execute the task.

```
C
C*** The number of equations is MM.
C*** SK is the stiffness matrix,
C*** A is the matrix of unknowns and
C*** F is the matrix of right hand sides.
C*** SK and F quantities are assumed to be in store.
C*** K,I and J are counters for elimination block,
C*** matrix row and matrix column.
C*** RM is the current multiplier.
C
      DIMENSION SK(500,500),A(500),F(500)
      DO 300 K=1,MM-1
      CALL ELIMINATE(SK,F,MM,K)
  300 CONTINUE
      A(MM)=F(MM)/SK(MM,MM)
      DO 700 JK=1,MM-1
      CALL BACKSUB(SK,F,A,MM,JK)
  700 CONTINUE
      STOP
      END
C
C
      SUBROUTINE ELIMINATE(SK,F,K,MM)
      DIMENSION SK(500,500),F(500)
      DO 300 I=K+1,MM
      RM=SK(I,K)/SK(K,K)
      SK(I,K)=0.
      DO 301 J=K+1,MM
  301 SK(I,J)=SK(I,J)-RM*SK(K,J)
  300 F(I)=F(I)-RM*F(K)
      RETURN
      END
C
C
C
      SUBROUTINE BACKSUB(SK,F,A,MM,JK)
      DIMENSION SK(500,500),F(500),A(500)
      I=MM-JK
      S=0.
      DO 800 L=I+1,MM
  800 S=S+F(L)*SKI,L)
      A(I)=(F(I)-S)/SK(I,I)
      RETURN
      END
```

There are many modifications to the basic elimination procedure which make it more efficient and economize on computer storage. Many **K** are symmetrical and then only half of **K** needs to be stored. Similarly, **K** may contain many zeros (i.e. is sparse) and these need not be stored. A consideration of the elimination process between (3) and (4) shows that the zero in the top right hand corner does not affect the elimination matrix, and banded solvers are available to make use of this fact. Some solvers, called frontal procedures, additionally deal with subsets of the equations and this improves efficiency. If the equation set is non-linear, the same general procedure suffices except that it must be used iteratively. The initial solution values are guessed so that **K** can be constructed and the resulting equations solved to give more precise **a**. The procedure can usually be repeated to convergence.

Given the presence of an equation solver, it is now important to answer questions about how **K** and **f** are calculated. The two principal methods are called *Finite Difference* and *Finite Element* procedures, and details of the schemes will now be discussed.

4. FINITE DIFFERENCE METHODS

4.1 General Procedures

The finite difference method deals directly with the underlying differential equations which describe the physical nature of the field problem. For example, if the field which is to be investigated is a temperature field, then the relevant differential equation, subject to suitable boundary conditions, is the transient heat flow equation

$$\frac{\partial^2 T}{\partial x^2} + \frac{\partial^2 T}{\partial y^2} + \frac{\partial^2 T}{\partial z^2} = \frac{1}{k}\frac{\partial T}{\partial t} \qquad (7)$$

where x, y and z are cartesian co-ordinates, t is time, T is temperature and k is the (assumed constant) thermal diffusivity. The finite difference procedure merely replaces the differential quantities by suitable differences and hence produces a set of equations for solution. The transfer from differential to difference is carried out through a suitable discretization process and the scheme outlined in Figure 2a is generally chosen. Attention is thus focused on the values of the field variables at the nodal points and these points are usually arranged on a regular equi-spaced grid, although other arrangements are possible. Suppose that three such adjacent nodes in the x direction have field variables a_{i-1}, a_i and a_{i+1} and that the nodes are equally spaced at separation h (Figure 3).

The first differential at the point i can then be represented by either a forward difference or a backward difference, i.e.

$$\left(\frac{\partial a}{\partial x}\right)_f \sim \frac{a_{i+1} - a_i}{h} \quad \text{(forward difference)}$$

$$\left(\frac{\partial a}{\partial x}\right)_b \sim \frac{a_i - a_{i-1}}{h} \quad \text{(backward difference)} \qquad (8)$$

104

These differences can be thought of as the approximations to the first differentials at points midway between $i-1$ and i and i and $i+1$ respectively, and a repeated application of the difference formulation will produce an estimate of the second differential. Thus

$$\frac{\partial^2 a}{\partial x^2} \sim \frac{(\frac{\partial a}{\partial x})_f - (\frac{\partial a}{\partial x})_b}{h^2}$$

$$= \frac{a_{i+1} - 2a_i + a_{i-1}}{h^2} \tag{9}$$

In view of its symmetric character, this is called a central difference. If other adjacent nodes are used, it is straightforward to write down approximations for higher differentials and the same general formulae apply to the y and z directions and, if necessary, with respect to time. Substitution of these formulae into equation (7) will result in the standard form $\mathbf{K}\,\mathbf{a} = \mathbf{f}$ as will be illustrated in the next section.

4.2 Boundary value problems

Consider a square two dimensional metal plate illustrated in Figure 4. Suppose that the left hand edge is maintained at 100°C and the right hand edge at 0°C. The top edge shows a linear variation of temperature but the variation along the bottom edge is approximately parabolic. If those temperatures are maintained constant, then the problem of establishing the internal temperatures is a boundary value problem and the governing differential equation when $\partial T/\partial t = 0$ can be written down from equation (7) as

$$\frac{\partial^2 T}{\partial x^2} + \frac{\partial^2 T}{\partial y^2} = 0 \tag{10}$$

Suppose that the plate is discretized by the square grid shown in Figure 4. Then the problem reduces to finding the temperatures at the nine internal nodes numbered $1 \to 9$. The finite difference equation is produced by replacing the differentials in equation (10) by central differences a node at a time.

Using equation (9) for node 1,

$$\frac{\partial^2 T}{\partial x^2} \sim \frac{1}{625}[100 - 2T_1 + T_2]$$

$$\frac{\partial^2 T}{\partial y^2} \sim \frac{1}{625}[75 - 2T_1 + T_4]$$

105

so that the differential equation obtained by adding together the differentials, as in equation (10), is:

$$4T_1 - T_2 - T_4 = 175 \tag{11}$$

This can be written in matrix form as

$$[4 \quad -1 \quad 0 \quad -1 \quad 0 \quad 0 \quad 0 \quad 0 \quad 0] \begin{bmatrix} T_1 \\ T_2 \\ T_3 \\ T_4 \\ T_5 \\ T_6 \\ T_7 \\ T_8 \\ T_9 \end{bmatrix} = [175] \tag{12}$$

In a similar way, the difference equations can be used at each of the nine internal nodes to develop nine linear equations.

$$\begin{bmatrix} 4 & -1 & 0 & -1 & 0 & 0 & 0 & 0 & 0 \\ -1 & 4 & -1 & 0 & -1 & 0 & 0 & 0 & 0 \\ 0 & -1 & 4 & 0 & 0 & -1 & 0 & 0 & 0 \\ -1 & 0 & 0 & 4 & -1 & 0 & -1 & 0 & 0 \\ 0 & -1 & 0 & -1 & 4 & -1 & 0 & -1 & 0 \\ 0 & 0 & -1 & 0 & -1 & 4 & 0 & 0 & -1 \\ 0 & 0 & 0 & -1 & 0 & 0 & 4 & -1 & 0 \\ 0 & 0 & 0 & 0 & -1 & 0 & -1 & 4 & -1 \\ 0 & 0 & 0 & 0 & 0 & -1 & 0 & -1 & 4 \end{bmatrix} \begin{bmatrix} T_1 \\ T_2 \\ T_3 \\ T_4 \\ T_5 \\ T_6 \\ T_7 \\ T_8 \\ T_9 \end{bmatrix} = \begin{bmatrix} 175 \\ 50 \\ 25 \\ 100 \\ 0 \\ 0 \\ 194 \\ 75 \\ 44 \end{bmatrix} \tag{13}$$

Even for the small number of nodes, the **K** matrix is strongly banded with many zeros so that a banded solver would be useful. As is usual for thermal problems, the matrix is also symmetric. Use of the simple solver given in section 3 yields

$$T_1 = 76.5 \quad T_2 = 52.1 \quad T_3 = 26.5 \quad T_4 = 78.9 \quad T_5 = 55.5$$
$$T_6 = 29.3 \quad T_7 = 83.7 \quad T_8 = 62.0 \quad T_9 = 33.7$$

A more precise solution could be obtained using a larger number of nodal points (i.e. smaller h) but this requires a small computer program to carry out the repetitive tasks.

4.3 Initial value problems

As an example of an initial value problem, consider a thin metal bar whose initial temperature is 0°C (Figure 5). If the temperature of the left hand end is suddenly changed to 100°C, whilst that of the right hand end is maintained at 0°C, then the temperatures of intermediate points will gradually rise to an equilibrium value. Determining the rate at which this occurs constitutes an initial value problem. The governing differential equation is derived from equation (7) and is

$$\frac{\partial^2 T}{\partial x^2} = \frac{1}{k}\frac{\partial T}{\partial t} \qquad (14)$$

where $T_{(x=0)} = 100°C$ and $T_{(x=100)} = 0°C$ for all t

$T_x = 0°C$ for $t = 0$.

The finite difference discretization for the x direction in the bar consists of nine equally spaced nodes (i.e. $h = 10mm$) and replacement of $\partial^2 T/\partial x^2$ in equation (14) can be made by the central difference method. We now need a replacement for the time derivative and several possibilities exist. The simplest is to use a forward difference in time. Let $T_{i,j}$ represent the temperature of the ith node at time j. Then if $T_{i,j+1}$ represents the temperature at one time step later, then

$$\frac{\partial T}{\partial t} \sim \frac{T_{i,j+1} - T_{i,j}}{g} \qquad (15)$$

where g is the size of the time step. We can now write down the finite difference form of equation (14). At position i and time j,

$$\frac{T_{i+1,j} - 2T_{i,j} + T_{i-1,j}}{h^2} = \frac{T_{i,j+1} - T_{i,j}}{kg}$$

Rearranging this equation gives an *explicit* form for $T_{i,j+1}$,

$$T_{i,j+1} = \frac{kg}{h^2}\{T_{i+1,j} - 2T_{i,j} + T_{i-1,j}\} + T_{i,j} \qquad (16)$$

The entire problem can thus be solved for as long a time as required without solving a linear equation. Suppose that for Figure 1, we have $h = 10$ mm, $k = 47$ mm²s⁻¹ and $g = 0.5$ s. Then for the initial

107

condition, $T_{1,0}$ to $T_{9,0}$ are all zero. Then, after 0.5 seconds, applying equation (16) to node 1,

$$T_{1,1} = \frac{47 \times 0.5}{100}(0 - 2 \times 0 + 100) + 0 \qquad = 23.5° C$$

For nodes 2 to 9,
$$T_{i,1} = \frac{47 \times 0.5}{100}(0 - 2 \times 0 + 0) + 0 \qquad = 0° C$$

The process is now repeated using the updated temperatures for a further time step (i.e. total time = 1.00)

$$T_{1,2} = \frac{47 \times 0.5}{100}(0 - 2 \times 23.5 + 100) + 23.5 \qquad = 40.0° C$$

$$T_{2,2} = \frac{47 \times 0.5}{100}(0 - 2 \times 0 + 23.5) + 0 \qquad = 5.5° C$$

and for i=3,9
$$T_{i,2} = \frac{47 \times 0.5}{100}(0 - 2 \times 0 + 1) + 0 \qquad = 0° C$$

Further time steps can now be taken, and the result is shown in Table 2.

The explicit scheme is very simple to program but suffers from the fact that many small time steps are required for accuracy. A detailed error analysis of the example given shows that the time step cannot exceed $h^2/2k$ if numerical stability is to be maintained, and this limit can often be very restrictive. The problem is alleviated if *implicit* schemes are used. These are derived in a very similar manner to the above except that a backward difference is used for the time differential. This results in a linear equation set at each time step and this is more time consuming than the explicit method. However, it is unconditionally stable, regardless of the size of g.

5. FINITE ELEMENT METHODS

5.1 General Procedures

Finite element procedures discretize the body using a scheme similar to that in Figure 2b. The individual discretized blocks are called elements and they are considered to be joined together at the nodes. In order to proceed with the approximation, the field variable is assumed to change in some simple way (e.g. linearly) across the element and the values at the nodes are the problem unknowns. As for the finite difference method, account has to be taken of the underlying physical equations governing the problem, but the method

108

by which this is done is different. Instead of substituting differences for derivatives, the basic partial differential equation is transformed into an integral form and this is used directly for the approximation. This transformation can be done in many ways. It is possible to make the change by a weighted residual approach to the differential equation or, alternatively, to seek some integral statement of the physics. Thus, for linear elastic examples, it is often sufficient to write down an expression for potential energy and then minimize it by the normal means of calculus. The whole procedure can be made much clearer by a boundary value example taken from plane elasticity theory.

5.2 Boundary Value Problems

Figure 6 shows a plane body subject to external forces. It is discretized with triangular elements. Some of the surface nodes are loaded by forces F and others are restrained to prevent rigid body motion. The field variables which are the unknowns are the nodal displacements and there are two of them for each node. This does not complicate the analysis: it merely doubles the number of unknowns in the final linear equation set.

Attention is now focused on a single element (Figure 7). The nodes of this element are identified by the numbers i, j, k and the numbering is conducted in an anticlockwise manner. Thus in Figure 7, $i<j<k$. Each node will have cartesian co-ordinates x and y and these are subscripted x_i, y_i etc., depending on the nodal number. In response to the applied loads, each node will undergo a displacement and this displacement will have components u and v parallel to the co-ordinate axes x and y. Again, these displacements will be subscripted according to the nodal number. It is convenient to gather all the displacements together as a 6×1 matrix δ where

$$\delta^T = [u_i \quad v_i \quad u_j \quad v_j \quad u_k \quad v_k] \tag{17}$$

It is also possible that the nodes are loaded by forces in the x and y directions. These forces will be designated FX and FY and will again be subscripted depending on their nodal number. They are conveniently grouped together into the 6×1 matrix $\mathbf{F}$ where

$$\mathbf{F}^T = [FX_i \quad FY_i \quad FX_j \quad FY_j \quad FX_k \quad FY_k] \tag{18}$$

The approximate values of the displacement within an element are approximated by a simple polynomial. The polynomial chosen must not allow gaps to develop between edges and the simplest functions are a combination of polynomials:-

$$u = \alpha_1 + \alpha_2 x + \alpha_3 y$$

$$v = \alpha_4 + \alpha_5 x + \alpha_6 y \tag{19}$$

where the α values are constants. It is possible to write the values of the constants, α from a knowledge of the nodal displacements. For instance

$$u_i = \alpha_1 + \alpha_2 x_i + \alpha_3 y_i$$

$$v_i = \alpha_4 + \alpha_5 x_i + \alpha_6 y_i \tag{20}$$

and there are four similar equations for the other two nodes. Retaining the matrix notation, equation (20) can be written as

$$
\begin{bmatrix} u_i \\ v_i \\ u_j \\ v_j \\ u_k \\ v_k \end{bmatrix} =
\begin{bmatrix}
1 & x_i & y_i & 0 & 0 & 0 \\
0 & 0 & 0 & 1 & x_i & y_i \\
1 & x_j & y_j & 0 & 0 & 0 \\
0 & 0 & 0 & 1 & x_j & y_j \\
1 & x_k & y_k & 0 & 0 & 0 \\
0 & 0 & 0 & 1 & x_k & y_k
\end{bmatrix}
\begin{bmatrix} \alpha_1 \\ \alpha_2 \\ \alpha_3 \\ \alpha_4 \\ \alpha_5 \\ \alpha_6 \end{bmatrix}
$$

i.e. $\delta = A \alpha$ (21)

Hence the polynomial coefficients can be obtained by inverting A to give

$$\alpha = A^{-1} \delta \tag{22}$$

Within the element, there are three strains ε_{xx}, ε_{yy} and ε_{xy}. The first two of these are normal strains and the last a shear strain. Using the definitions of small strain theory, the strain matrix, ε is

$$
\varepsilon = \begin{bmatrix} \varepsilon_{xx} \\ \varepsilon_{yy} \\ \varepsilon_{xy} \end{bmatrix} =
\begin{bmatrix} \partial u / \partial x \\ \partial v / \partial y \\ \partial u / \partial y + \partial v / \partial x \end{bmatrix} \tag{23}
$$

Thus, in terms of equation (19),

$$\varepsilon_{xx} = \alpha_2 , \quad \varepsilon_{yy} = \alpha_6 \text{ and } \varepsilon_{xy} = \alpha_3 + \alpha_5 \tag{24}$$

Since none of the expressions in equation (24) contain x and y explicitly, the strains within the element are constant with position (as, therefore, are the stresses). In matrix notation,

110

$$\varepsilon = \begin{bmatrix} \varepsilon_{xx} \\ \varepsilon_{yy} \\ \varepsilon_{xy} \end{bmatrix} = \begin{bmatrix} 0 & 1 & 0 & 0 & 0 & 0 \\ 0 & 0 & 0 & 0 & 0 & 1 \\ 0 & 0 & 1 & 0 & 1 & 0 \end{bmatrix} \begin{bmatrix} \alpha_1 \\ \alpha_2 \\ \alpha_3 \\ \alpha_4 \\ \alpha_5 \\ \alpha_6 \end{bmatrix}$$

i.e. $\varepsilon = C \, \alpha$ (25)

Remembering equation (22)

$$\varepsilon = C \, A^{-1} \, \delta$$

or $\varepsilon = B \, \delta$ (26)

where $B = C \, A^{-1}$

with a little algebraic effort, B can be obtained explicitly.

$$B = \frac{1}{2\Delta} \begin{bmatrix} y_j - y_k & 0 & y_k - y_i & 0 & y_i - y_j & 0 \\ 0 & x_k - x_j & 0 & x_i - x_k & 0 & x_j - x_i \\ x_k - x_j & y_j - y_k & x_i - x_k & y_k - y_i & x_j - x_i & y_i - y_j \end{bmatrix} \quad (27)$$

where Δ is the area of the triangular element.

Corresponding to the strain matrix, ε, there will be a 3×1 stress matrix, σ. These two matrices are linearly related through the generalized Hooke's Law for plane stress conditions.

$$\sigma = \begin{bmatrix} \sigma_{xx} \\ \sigma_{yy} \\ \sigma_{xy} \end{bmatrix} = \frac{E}{1 - \upsilon^2} \begin{bmatrix} 1 & \upsilon & 0 \\ \upsilon & 1 & 0 \\ 0 & 0 & (1-\upsilon)/2 \end{bmatrix} \begin{bmatrix} \varepsilon_{xx} \\ \varepsilon_{yy} \\ \varepsilon_{xy} \end{bmatrix} \quad (28)$$

i.e. $\sigma = D \, \varepsilon$

where E and υ are Young's modulus and Poisson's ratio respectively. Thus, in terms of nodal displacements,

$$\sigma = D \, B \, \delta \quad (29)$$

At this stage, the physical principles which underlie the problem must be brought into play. In the present case the potential energy of the system, χ, must take on a minimum value for equilibrium. Now

χ = Strain energy stored + Potential energy of external forces

Suppose that the displacements for all the nodes in the element change by a small amount $d\delta$ (this is a 6×1 matrix corresponding

111

to δ). Then the corresponding change in χ, $d\chi$ is

$$d\chi = \int_{\Delta} (\sigma_{xx} d\varepsilon_{xx} + \sigma_{yy} d\varepsilon_{yy} + \sigma_{xy} d\varepsilon_{xy}) d\Delta - F^T d\delta \qquad (30)$$

where the integration is carried out over the area, Δ of the element. Since the element stresses and strains are independent of the co-ordinates in the element, the integral is particularly easy to evaluate. Hence

$$d\chi = \sigma^{\mathsf{T}} d\varepsilon - F^{\mathsf{T}} d\delta \qquad (31)$$

Using the equations already derived from strain (equation 26)

$$d\chi = \sigma^{\mathsf{T}} B \, d\delta - F^{\mathsf{T}} d\delta \qquad (32)$$

For χ to be a minimum, $d\chi = 0$ for arbitrary $d\delta$ i.e.

$$\Delta \sigma^{\mathsf{T}} B = F^{\mathsf{T}}$$

Transposing gives

$$\Delta B^{\mathsf{T}} \sigma = F$$

and remembering the expression for σ (equation 29)

$$\Delta B^{\mathsf{T}} D B \, \delta = F \qquad (33)$$

This is a set of linear equations in the six unknown quantities in δ with all the other quantities known (i.e. nodal co-ordinates, forces, etc.). The 6×6 matrix on the left hand side of the equation is known as the element stiffness matrix and is symmetric.

Equations of the form of (33) can be obtained for each element in the structure. The equations can then be assembled by adding the suitably arranged matrices together to give an overall stiffness matrix and a linear equation set in all the nodal displacements in the body. When the displacements have been obtained by solution of equation (33), the quantities stress and strain can be derived from them at an elemental level by recalling the already calculated **B** and **D** matrices and using equations (26) and (29).

The following simple example can be carried through by hand calculation and this process will make clear the theoretical analysis outlined above. A square rectangular plate of dimension 1×1 is loaded at two corners by loads of 100 and 200 respectively (Figure 8). It is required to calculate the displacements of the structure as well as the stresses through the body. The material has a Young's modulus of 10,000 and a Poisson's ratio of 1/3. The plate is divided into two triangular elements I and II with four nodal points. These are numbered 1 to 4. Nodes 3 and 4 are externally loaded by loads of

200 and 100 respectively in the y direction. Nodes 1 and 2 are restrained from movement as shown in Figure 8. The finite element calculations are now conducted element by element before assembly and final solution.

ELEMENT I

	x	y	u	v	FX	FY	
i	0	0	u_1	v_1	0	0	E = 10,000
j	1	0	u_2	v_2	0	0	v = 1/3
k	1	1	u_3	v_3	0	0	

$$D = 11250 \begin{bmatrix} 1 & 0.33 & 0 \\ 0.33 & 1 & 0 \\ 0 & 0 & 0.33 \end{bmatrix}$$

$$B = \frac{1}{2 \times 0.5} \begin{bmatrix} -1 & 0 & 1 & 0 & 0 & 0 \\ 0 & 0 & 0 & -1 & 0 & 1 \\ 0 & -1 & -1 & 1 & 1 & 0 \end{bmatrix}$$

$$B^T D = 11250 \begin{bmatrix} -1 & -0.33 & 0 \\ 0 & 0 & -0.33 \\ 1 & 0.33 & -0.33 \\ -0.33 & -1 & 0.33 \\ 0 & 0 & 0.33 \\ 0.33 & 1 & 0 \end{bmatrix}$$

$$B^T D B = 11250 \begin{bmatrix} 1 & 0 & -1 & 0.33 & 0 & -0.33 \\ 0 & 0.33 & 0.33 & -0.33 & -0.33 & 0 \\ -1 & 0.33 & 1.33 & -0.66 & -0.33 & 0.33 \\ 0.33 & -0.33 & -0.66 & 1.33 & 0.33 & -1 \\ 0 & -0.33 & -0.33 & 0.33 & 0.33 & 0 \\ -0.33 & 0 & 0.33 & -1 & 0 & 1 \end{bmatrix}$$

$$D B = 11250 \begin{bmatrix} -1 & 0 & 1 & -0.33 & 0 & 0.33 \\ -0.33 & 0 & 0.33 & -1 & 0 & 1 \\ 0 & -0.33 & -0.33 & 0.33 & 0.33 & 0 \end{bmatrix}$$

113

$$5625 \begin{bmatrix} 1 & 0 & -1 & 0.33 & 0 & 0.33 \\ & 0.33 & 0.33 & -0.33 & -0.33 & 0 \\ & & 1.33 & -0.66 & -0.33 & 0.33 \\ & & & 1.33 & 0.33 & -1 \\ & \text{Sym} & & & 0.33 & 0 \\ & & & & & 1 \end{bmatrix} \begin{bmatrix} u_1 \\ v_1 \\ u_2 \\ v_2 \\ u_3 \\ v_3 \end{bmatrix} = \begin{bmatrix} 0 \\ 0 \\ 0 \\ 0 \\ 0 \\ 0 \end{bmatrix}$$

ELEMENT II

	x	y	u	v	FX	FY
i	0	0	u_1	v_1	0	0
j	1	1	u_3	v_3	0	-200
k	0	1	u_4	v_4	0	0-100

$E = 10,000$
$v = 1/3$

$$\mathbf{D} = 11250 \begin{bmatrix} 1 & 0.33 & 0 \\ 0.33 & 1 & 0 \\ 0 & 0 & 0.33 \end{bmatrix}$$

$$\mathbf{B} = \frac{1}{2 \times 0.5} \begin{bmatrix} 0 & 0 & 1 & 0 & -1 & 0 \\ 0 & -1 & 0 & 0 & 0 & 1 \\ -1 & 0 & 0 & 1 & 1 & -1 \end{bmatrix}$$

$$\mathbf{B^T D} = \begin{bmatrix} 0 & 0 & -0.33 \\ -0.33 & -1 & 0 \\ 1 & 0.33 & 0 \\ 0 & 0 & 0.33 \\ -1 & -0.33 & 0.33 \\ 0.33 & 1 & -0.33 \end{bmatrix}$$

$$\mathbf{B^T D B} = 11250 \begin{bmatrix} 0.33 & 0 & 0 & -0.33 & -0.33 & 0.33 \\ 0 & 1 & -0.33 & 0 & 0.33 & -1 \\ 0 & -0.33 & 1 & 0 & -1 & 0.33 \\ -0.33 & 0 & 0 & 0.33 & 0.33 & -0.33 \\ -0.33 & 0.33 & -1 & 0.33 & 1.33 & -0.66 \\ -0.33 & 0 & 0.33 & -1 & 0 & 1 \end{bmatrix}$$

114

$$\mathbf{DB} = 11250 \begin{bmatrix} 0 & -0.33 & 1 & 0 & -1 & 0.33 \\ 0 & -1 & 0.33 & 0 & -0.33 & 1 \\ -0.33 & 0 & 0 & 0.33 & 0.33 & -0.33 \end{bmatrix}$$

$$5625 \begin{bmatrix} 0.33 & 0 & 0 & -0.33 & -0.33 & 0.33 \\ & 1 & -0.33 & 0 & 0.33 & -1 \\ & & 1 & 0 & -1 & 0.33 \\ & & & 0.33 & 0.33 & -0.33 \\ \text{Sym} & & & & 1.33 & -0.66 \\ & & & & & 1.33 \end{bmatrix} \begin{bmatrix} u_1 \\ v_1 \\ u_3 \\ v_3 \\ u_4 \\ v_4 \end{bmatrix} = \begin{bmatrix} 0 \\ 0 \\ 0 \\ -200 \\ 0 \\ -100 \end{bmatrix}$$

In order to make the process of element assembly clear, the two element stiffness matrices are now written in expanded form with relation to the overall 8 field variables.

$$5625 \begin{bmatrix} 1 & 0 & -1 & 0.33 & 0 & -0.33 & - & - \\ 0 & 0.33 & 0.33 & -0.33 & -0.33 & 0 & - & - \\ -1 & 0.33 & 1.33 & -0.66 & -0.33 & 0.33 & - & - \\ 0.33 & -0.33 & -0.66 & 1.33 & 0.33 & -1 & - & - \\ 0 & -0.33 & -0.33 & 0.33 & 0.33 & 0 & - & - \\ -0.33 & 0 & 0.33 & -1 & 0 & 1 & - & - \\ - & - & - & - & - & - & - & - \\ - & - & - & - & - & - & - & - \end{bmatrix} \begin{bmatrix} u_1 \\ v_1 \\ u_2 \\ v_2 \\ u_3 \\ v_3 \\ u_4 \\ v_4 \end{bmatrix} = \begin{bmatrix} 0 \\ 0 \\ 0 \\ 0 \\ 0 \\ 0 \\ - \\ - \end{bmatrix}$$

$$+ \qquad\qquad\qquad +$$

$$5625 \begin{bmatrix} 0.33 & 0 & - & - & 0 & -0.33 & -0.33 & 0.33 \\ 0 & 1 & - & - & -0.33 & 0 & 0.33 & -1 \\ - & - & - & - & - & - & - & 0 \\ - & - & - & - & - & - & 0 & - \\ 0 & -0.33 & - & - & 1 & 0 & -1 & 0.33 \\ -0.33 & 0 & - & - & 0 & 0.33 & 0.33 & -0.33 \\ -0.33 & 0.33 & - & - & -1 & 0.33 & 1.33 & -0.66 \\ 0.33 & -1 & - & - & 0.33 & -0.33 & -0.66 & 1.33 \end{bmatrix} \begin{bmatrix} u_1 \\ v_1 \\ u_2 \\ v_2 \\ u_3 \\ v_3 \\ u_4 \\ v_4 \end{bmatrix} = \begin{bmatrix} 0 \\ 0 \\ - \\ - \\ 0 \\ -200 \\ 0 \\ -100 \end{bmatrix}$$

These are assembled together as follows.

$$5625 \begin{bmatrix} 1.33 & 0 & -1 & 0.33 & 0 & -0.66 & -0.33 & 0.33 \\ 0 & 1.33 & 0.33 & -0.33 & -0.66 & 0 & 0.33 & -1 \\ -1 & 0.33 & 1.33 & -0.66 & -0.33 & 0.33 & 0 & 0 \\ 0.33 & -0.33 & -0.66 & 1.33 & 0.33 & -1 & 0 & 0 \\ 0 & -0.66 & -0.33 & 0.33 & 1.33 & 0 & -1 & 0.33 \\ -0.66 & 0 & 0.33 & -1 & 0 & 1.33 & 0.33 & -0.33 \\ -0.33 & 0.33 & 0 & 0 & -1 & 0.33 & 1.33 & 0.66 \\ 0.33 & -1 & 0 & 0 & 0.33 & -0.33 & -0.66 & 1.33 \end{bmatrix} \begin{bmatrix} u_1 \\ v_1 \\ u_2 \\ v_2 \\ u_3 \\ v_3 \\ u_4 \\ v_4 \end{bmatrix} = \begin{bmatrix} 0 \\ 0 \\ 0 \\ 0 \\ 0 \\ -200 \\ 0 \\ -100 \end{bmatrix}$$

Since the displacements u_1, v_1 and v_2 are set to zero to avoid rigid body rotation, these variables can be eliminated from the equation set resulting in the following 5 equation linear set. The solution can now be obtained by Gauss elimination.

$$5625 \underset{\mathbf{k}}{\begin{bmatrix} 1.33 & -0.33 & 0.33 & 0 & 0 \\ -0.33 & 1.33 & 0 & -1 & 0.33 \\ 0.33 & 0 & 1.33 & 0.33 & -0.33 \\ 0 & -1 & 0.33 & 1.33 & -0.66 \\ 0 & 0.33 & -0.33 & -0.66 & 1.33 \end{bmatrix}} \underset{\mathbf{a}}{\begin{bmatrix} u_2 \\ u_3 \\ v_3 \\ u_4 \\ v_4 \end{bmatrix}} = \underset{\mathbf{f}}{\begin{bmatrix} 0 \\ 0 \\ -200 \\ 0 \\ -100 \end{bmatrix}} \quad \underset{\text{solution}}{\begin{bmatrix} u_2 \\ u_3 \\ v_3 \\ u_4 \\ v_4 \end{bmatrix} = \begin{bmatrix} 0.0145 \\ 0.0189 \\ -0.0385 \\ 0.0138 \\ -0.0211 \end{bmatrix}}$$

Given the nodal displacements for each element, the stresses can be calculated using equations (26) and (29) as follows.

Element I

$$\delta = \begin{bmatrix} 0 \\ 0 \\ 0.0145 \\ 0 \\ 0.0189 \\ -0.0385 \end{bmatrix} \quad \sigma = \begin{bmatrix} 16 \\ -383 \\ 18 \end{bmatrix}$$

Element II

$$\delta = \begin{bmatrix} 0 \\ 0 \\ 0.0189 \\ -0.0385 \\ 0.0133 \\ -0.0211 \end{bmatrix} \quad \sigma = \begin{bmatrix} -26 \\ -215 \\ -16 \end{bmatrix}$$

5.3 Initial Value Problems

Many initial value problems have been solved by the finite element procedure. The extension is quite straightforward. The field variables to be determined are taken to be the time gradients and these are updated with respect to time by exactly the same procedures as for as the explicit and implicit finite difference techniques. Thus the procedures can be thought of as a combination of finite difference and finite element methods. As an example, flow problems may proceed by finding $\partial u/\partial t$ and $\partial v/\partial t$ for all nodal positions and then updating them by a forward difference so that

116

$$u(t = 1) = u(t = 0) + \left(\frac{\partial u}{\partial t}\right) \times g$$

$$\text{and} \quad v(t = 1) = u(t = 0) + \left(\frac{\partial v}{\partial t}\right) \times g \tag{34}$$

6. GENERAL EXTENSIONS

Previous sections have indicated the general methodology of commonly used numerical methods in the area of material technology. Once these basic ideas have been grasped, it is relatively straightforward to extend them to a whole range of quite complex problems with only minor alterations to the general procedures. This section will list some of these extensions but is by no means exhaustive.

6.1 Elastic-Plastic Analysis

In many areas of metallurgical interest, materials do not remain elastic on loading but yielding and plasticity occur. Figure 9 shows a typical stress-strain curve plotted in terms of effective stress $\bar{\sigma}$ and effective strain $\bar{\varepsilon}$. *

As the stress is gradually increased from zero, behaviour is linearly elastic up to the stress $\bar{\sigma}_y$. Following this, behaviour is non-linear and unrecoverable so that the one to one relationship between stress and strain is destroyed. Any modelling of this process must therefore be incremental and finite element procedures work by incrementally increasing loads and following the gradual change, element by element, in displacement stresses and strains. Since the slope of the plastic stress-strain curve, H (Figure 9) changes with stress and strain, different elements in the structure have different properties. What is required is to redefine the $\mathbf{D}$ matrix of equation (28) in a suitable form for incremental loading. The incremental strain $\mathbf{d\varepsilon}$ can be split into elastic and plastic constituents

$$\mathbf{d\varepsilon} = \mathbf{d\varepsilon_e} + \mathbf{d\varepsilon_p} \tag{35}$$

Now $\mathbf{d\varepsilon_e} = \mathbf{D}\,\mathbf{d\sigma}$ so that $\mathbf{d\sigma} = \mathbf{D^{-1}}(\mathbf{d\varepsilon} - \mathbf{d\varepsilon_p})$ and thus can be written as

$$\mathbf{d\sigma} = \mathbf{C}\,\mathbf{d\varepsilon},$$
$$\mathbf{d\varepsilon} = \mathbf{C^{-1}}\mathbf{d\sigma} \tag{36}$$

$\mathbf{C^{-1}}$ is the elastic-plastic matrix which replaces $\mathbf{D}$ in equation (28) for those elements which have yielded. It can be shown to be a function of the current deviatoric stress field and the slope of the stress-strain curve, H. The finite element formulation is now very similar to

*Definition of these generalised stresses and strains can be found in Advanced Mechanics of Materials, H Ford and J M Alexander, John Wiley, New York and Chichester, 1977.

117

section 5 except that the stiffness matrix **K** is now a function of the current element stress and strain. The eventual equation set is thus non-linear but this does not present any special problems in solution.

6.2 Elastic-Creep Problems

In high temperature metallurgical plant the problem is not generally one of plastic yield but of slow time dependent deformation (creep) after initial elastic loading. This can give rise to considerable stress redistribution and possibly to failure in low ductility regions. Again, the finite element procedure outlined in section 5 can be modified to calculate such creep behaviour provided a relationship relating creep strain to stress and time is known. Let $\dot{\varepsilon} = f(\sigma)$ where f can be any suitable (often experimentally determined) function. The total strain ε at any time can now be split into elastic and creep strains

$$\varepsilon = \varepsilon_e + \varepsilon_c \tag{37}$$

As for section 6.1,

$$\sigma = D \varepsilon_e = D \varepsilon - D \varepsilon_c \tag{38}$$

and substitution into the potential energy equation (30) yields

$$\Delta \, B^T \, D \, B \, S = F - F_c \tag{39}$$

where F_c is the creep loading defined as $\Delta \, B^T \, D \, \varepsilon_c$. The problem is thus essentially similar to the elastic procedure except that the loading F is incremented by the creep loading. This involves the creep strain which is simply calculated as

$$\varepsilon_c^{(i+1)} = \varepsilon_c^{(i)} + f(\sigma,t)g \tag{40}$$

where the superscripts indicate updated quantities.

6.3 Flow (Visco-Plastic) Analysis

In many metalworking processes, the strains involved are so large that it is possible to ignore elastic effects and simply formulate a finite element procedure on the basis of plastic flow. The basic constitutive relationship is then written in terms of the Levy-Mises equations so that strain rates are related to deviatoric stresses

$$\dot{\varepsilon}_{ij} = \frac{\bar{\dot{\varepsilon}}}{\bar{\sigma}} \sigma'_{ij} \tag{41}$$

This implies the re-definition of **D** as (for the two dimensional case)

$$D = \frac{2\bar{\sigma}}{3\bar{\dot{\varepsilon}}} \begin{bmatrix} 1 & 0 & 0 \\ 0 & 1 & 0 \\ 0 & 0 & 1 \end{bmatrix} \tag{42}$$

118

with the replacement of σ by $\bar{\sigma}$ and ε by $\dot{\varepsilon}$. There is now a complete analogy between the elastic case and the visco-plastic case when the field variable is taken to be nodal velocities rather than nodal displacements. The equation set **K a = f** is now non-linear since **D** depends on current $\bar{\sigma}$ and $\bar{\varepsilon}$ but once a solution is obtained, nodal positions can be updated by means of the velocities and a suitable time step and the process repeated. This technique is particularly useful for large strain metal working operations.

6.4 Coupled Problems

A whole series of numerical procedures can be grouped together to tackle complicated coupled problems. A good example is the use of numerical procedures in the analysis of hot deformation. In general, the flow rates (visco-plastic analysis) will be strongly dependent on current temperature. However, temperature is affected by adiabatic heating and subsequent thermal diffusion and these temperature changes can be calculated by finite difference or element solvers. The changes in temperature will now affect the flow properties which will in turn affect the temperature changes. The whole process is strongly coupled, but a combination of techniques can deal with analysis. In a similar way, casting problems can be accurately modelled and such modelling can also include the development of metallurgical structure.

6.5 Fracture Mechanics

Finite elements can also be used to estimate stress concentration factors and J integrals at cracks in continuua. The difficulty here is not with the finite element formulation: standard linear elastic or elastic-plastic codes will suffice. It is rather with the geometry at the crack tip since the stress fields may well exhibit singularities here. Rather than resorting to very fine element sizes at such points, special 'crack-tip' elements have been derived with the same type of polynomial functions as the stress field is likely to exhibit.

7. CONCLUSIONS

The present article has been intended to provide the reader with an initial insight into numerical procedures in field problems. For those who are interested in further details there are many textbooks available with varying degrees of complexity. The author has found three books to be particularly useful.[1,2,3] The first two describe the general methods employed whilst the third gives an excellent insight into the problems involved with programming. However, many software packages are now available which will carry out several of the processes discussed in the article. It is not necessary to own a large computer or to understand in greater detail than here described the finite element procedures. One excellent software package[4] known to the author runs programs directly on P.C. machines and will deal with standard elastic, elastic-plastic, elastic-viscoplastic and thermal problems.

8. REFERENCES

1) AMES, W.F., 'Numerical Methods for Partial Differential Equations', Academic Press, New York.

2) DESAI, C.S. and ABEL, J.F., 'Introduction to the Finite Element Method', Van Nostrand Reinhold Company, New York.

3) HINTON, E. and OWEN, D.R.J., 'Finite Element Programming', Academic Press, London.

4) Rockfield Software, Innovation Centre, University College of Swansea, Swansea, U.K.

Table 1. Numerical Estimation of π.

Number of sides	Inscribed polygon	Exscribed polygon	Difference
3	2.598	5.196	2.598
4	2.828	4.000	1.712
8	3.062	3.314	0.252
16	3.121	3.183	0.0611
32	3.127	3.152	0.0152
64	3.140	3.144	0.004
128	3.141	3.142	0.001

Table 2. Explicit Finite Difference Scheme for Figure 5.

Time Step j	Total Time s	Nodal Temperatures °C								
		1	2	3	4	5	6	7	8	9
1	0.5	23.5	0	0	0	0	0	0	0	0
2	1.0	40.0	5.5	0	0	0	0	0	0	0
3	1.5	49.4	11.4	1.3	0	0	0	0	0	0
:										
10	5.0	65.4	36.8	17.6	7.0	2.2	0.6	0.1	0	0
:										
25	12.5	77.3	56.4	38.6	24.8	14.8	8.2	4.2	2.0	0.8
:										
50	25.0	83.7	68.1	53.8	41.1	30.3	21.4	14.3	8.7	4.0
:										
100	50.0	88.1	76.4	65.0	54.1	43.8	34.1	25.0	16.4	8.1

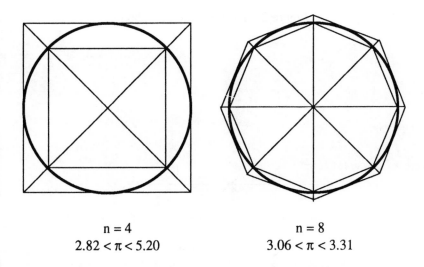

$n = 4$ $n = 8$

$2.82 < \pi < 5.20$ $3.06 < \pi < 3.31$

Figure 1 Approximating π by inscribed and exscribed polygons.

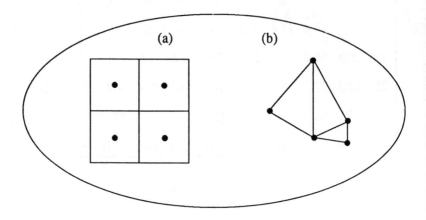

Figure 2   Discretization of a body.

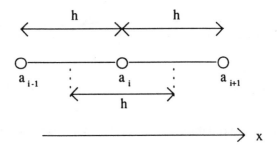

Figure 3   Finite difference nodes.

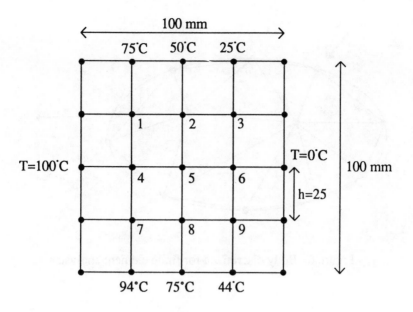

Figure 4   Finite difference discretization of a plate.

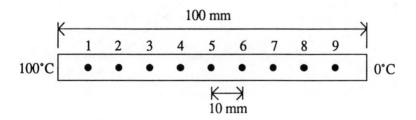

Figure 5   Discretization of a bar.

123

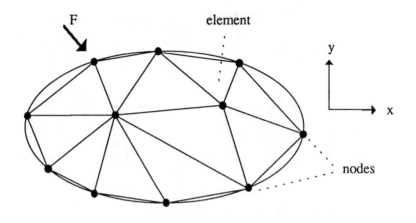

Figure 6   Body discretized for finite element analysis

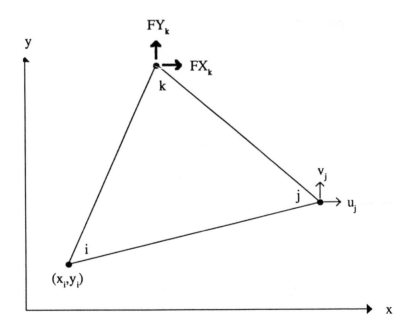

Figure 7   Single element

124

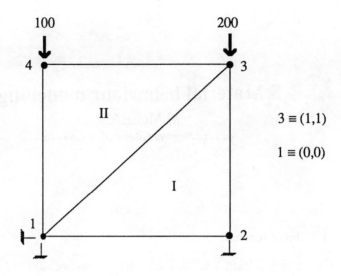

Figure 8   Plane stress plate

$3 \equiv (1,1)$

$1 \equiv (0,0)$

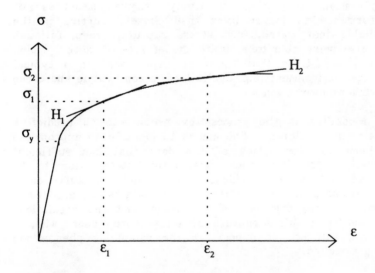

Figure 9   Effective stress/effective strain curve.

125

5:Material behaviour modelling

M McLEAN

National Physical Laboratory, Teddington

1   Introduction

It is unrealistic to expect that measurement of material properties can be made to cover all eventualities that would be required to represent the performance of the material in potential service conditions.   Rather, there must be a rational basis for interpolating or extrapolating from a more limited database with some assurance that this gives a realistic representation of the material response. In the case of high temperature deformation and fracture this does not only mean predicting relatively simple measures of performance to longer lives than those occurring in the available test matrix (e.g. stress rupture, creep, fatigue), but also being able to estimate the effects of more complex loading conditions that might be experienced in a typical service environment (e.g. cyclic load, creep-fatigue and creep corrosion environments.)

Modelling is the process by which such a predictive scheme is developed. The aim is to identify an appropriate mathematical description of the data that has sufficient flexibility to allow the extrapolation/interpolation to be carried out with confidence, but which is sufficiently general and simple to operate that the data can be encoded in terms of the formalism of the model and the predictions decoded from this structure by all potential users who, in general, will not be experts in the particular phenomenon being modelled.

There are two broad approaches to the modelling of mechanical properties (and indeed to all materials modelling) that impact on different objectives of the materials technologist and engineer. Models framed in terms of the microstructure of the material provide an invaluable guide to alloy and process development, identifying the key factors that must be controlled in order to optimize the material performance. On the other hand, prediction of service performance is facilitated by models that are expressed through engineering measures of the mechanical behaviour. Both are important and, indeed, it is possible to have two versions of the same model structured for each objective.

The detailed models of mechanical performance have been influenced by the computational facilities that have been available to both the research workers who have developed the models and to the engineers and materials technologists who use the models in a wide range of applications; widespread availability of computer power has revolutionised our approach to modelling. It is not possible to review the full scope of mechanical behaviour modelling in the space available. However, in this paper the nature of the various developments will be discussed through selected examples showing how they might contribute to:

- materials development
- engineering design
- remanent life assessment
- database compilation

2 Categories of models

Modelling of mechanical behaviour (and indeed to all modelling) falls into two general categories, and each has particular benefits.

a) Empirical models take a pragmatic view of the organisation of data. They look for mathematical patterns in the data that are available and assume that this description can be extended to arbitrary conditions for which there are no data. The model can be either framed in terms of simple measures of the mechanical behaviour (e.g. rupture life, minimum creep rate, fatigue life) or it can incorporate known features of the microstructure (e.g grain size, dislocation density).

127

The important point is that the model is developed through a process of hypothesis and validation - i.e. it is essentially an inductive process.

b) Physical models build on an understanding of the mechanisms that are known to control the particular phenomena under consideration. They are developed by a deductive process from a series of facts that are known (or believed) to be true and generally describe the phenomena in relation to the critical aspects of the microstructure of the materials.

The most appropriate model formulation depends very much on the application.

Materials development requires information on which features of the constitution or structure of the material must be modified. Thus microstructure-based models are required that provide guide-lines for, for example, solution or particle hardening or composite strengthening.

Database development for engineering design on an optimized material is most effectively aided by models framed in terms of operational parameters describing the form of the available data and allowing the database matrix to be minimised, particularly with reference to long term testing.

Engineering design will generally need to operate on an existing database, expressed in terms of operational rather than physical parameters, and to extend it to represent more complex operating environments. Consequently, models should have in-built flexibility to allow the effects of, for example, multiaxial stressing or concurrent corrosion to be taken into account.

Remanent life procedures can take different forms which require different types of models. If the evolution of a life-limiting defect is known then a physics-based model in conjunction with an appropriate inspection technique is required. However, if remaining life is determined by careful monitoring of the service conditions and comparison with the design life then an empirical description of the data is best.

128

The remainder of this paper will indicate how modelling of two important aspects of mechanical behaviour, creep and fatigue, have evolved in relation to the increasing availability of high capacity computing resources for both data analysis and design/remnant life calculations.

3   Creep

The deformation mechanism maps (Figure 1) popularised by Ashby[1] graphically illustrate that the dominant mechanisms of deformation vary with stress and temperature. Consequently, in developing a model of creep deformation it must always be realised that it can only be appropriate within a certain range of operating conditions and the fields of applicability of various models are quite different for different types of materials. Indeed, it is this author's view that the application of concepts that have been rigorously established for simple metals to complex engineering alloys is quite inappropriate and leads to serious problems in attempting to extrapolate data.

Figure 2 shows schematic creep curves that are characteristic of (a) simple single-phase metals and (b) complex engineering alloys such as nickel-base superalloys which are precipitation strengthened and are used in gas turbines. These typify the important differences between the creep behaviours of simple single-phase metals, on which most mechanistic studies have been carried out and on which the formalisms for representing creep data have been based, and real engineering materials. Here we shall discuss two aspects of the differences in creep behaviour, viz. the shapes of the creep curves and the equations describing characteristic creep rates.

(a)   The shapes of the creep curves are quite different. The simple metal has a relatively short primary region (decreasing creep rate) that is associated with hardening of the material due to increasing densities of dislocations and a short tertiary region (sharply increasing creep rate) that is the result of the generation and growth of damage (e.g. creep cavities). However, most of the creep life is characterised by a constant, or steady-state creep rate where there is believed to be a dynamic balance between new dislocations being generated during glide and dislocations being removed by coarsening (or recovery). Models

129

developed by D. McLean[2] and Lagneborg[3] in particular have been very successful in describing this type of behaviour.

In the case of engineering alloys there is normally no such extensive steady state behaviour: rather, after a short primary creep region most of the life is dominated by a progressively increasing, or tertiary, creep behaviour that is not associated with any obvious development of cavitation damage. Instead of a steady state creep rate there is an inflexion at which the creep rate takes a minimum value. Dyson and McLean[4] have modelled creep in such engineering alloys in terms of a progressive increase in mobile dislocation density rather than through the establishment of a steady state structure.

b) The steady state creep rate $\dot{\epsilon}_{ss}$ for the simple metals and the minimum creep rates $\dot{\epsilon}_{min}$ for engineering alloys vary over many orders of magnitude with relatively small variations in stress σ and temperature T as indicated in Figure 3. These data are often represented by power-law expressions:[7]

$$\dot{\epsilon}_{ss} = A\,\sigma^n \exp(-Q/RT) \qquad (1)$$

where A, n and Q are constants; R is the gas constant.

For simple single-phase alloys the representation requires values of $n \sim 4$ and $Q \sim$ activation energy for self-diffusion which are consistent with recovery-creep models of D. McLean[2] and Lagneborg.[3] However, for complex particle strengthened materials the values of $n \gg 4$ and $Q \gg Q_{SD}$ have no obvious physical significance. Of course, Equation 1 can always be used to map such data, and indeed it is extensively used for this purpose; however because there is no fundamental basis for doing so care must be taken in extrapolating beyond the range of the database.

Some attempts have been made to rationalise the inconsistencies between the recovery-creep model and the parameters of Equation 1 by introducing the concept of a threshold, or friction, stress σ_i.[8]

$$\dot{\epsilon}_{min} = A'\,(\sigma - \sigma_i)^n \exp(-Q/RT) \qquad (2)$$

This approach still has its advocates and there is no doubt that solutes, precipitates and other particles present obstacles to dislocation motion. However, it is difficult to measure the associated thresholds unambiguously. Of more importance, it depends on the basic assumption, without proof, that the deformation mechanism is unaltered by the strengthening elements.

As discussed above, the power-law (or Bailey-Norton) representation of characteristic creep rate is extensively used to organise much creep data. Other creep data are found to be better represented by an exponential law.

$$\dot{\epsilon}_{ss} = A \exp \left[\frac{\sigma}{\sigma_0} - \frac{Q}{RT} \right] \qquad (3)$$

Equation 3 can, indeed, be regarded as an approximation that is valid at high stresses for the equation:

$$\dot{\epsilon}_{ss} = 2A \sinh \left(\frac{\sigma}{\sigma_0} \right) \exp \left(-Q/RT \right) \qquad (4)$$

These expressions are consistent with physical models of creep which assume glide to be diffusion controlled and consider creep in terms of reaction-rate theory, rather than of recovery control.[7]

Various parametric representations of stress rupture data (and of the stress as a function of time to attain arbitrary strains) are extensively used by engineers to take maximum advantage of data obtained over a limited range of stresses and temperatures. These are of particular importance in design codes. The parameters which combine the contributions of stress and temperature provide a single master curve of the data. When the rupture life (or time to achieve specific strains) scales inversely with minimum creep rate, as they often do, various parametric representations that are widely used have at least an indirect relationship to the creep mechanisms described above. The Monkman-Grant[9] relationship which notes a constancy in the product of $\dot{\epsilon}_{ss} t_{ff}$ is satisfied by many materials (Figure 4) and the value of the constant of proportionality can be a powerful indicator of the operative mechanism.

The Larson-Miller[10] parametric representation is perhaps the most widely used approach to representing stress rupture data. Its validity is based on the premise that the equation describing t_f can be expressed as separable functions of stress and temperature and, consequently, is consistent with either of Equations 2 or 3.

131

$$t_f = A\ f(\sigma)\ \exp\ (Q/RT) \tag{5}$$

This can be expressed as:

$$\log t_f - \log A = \log f(\sigma) + \frac{Q}{RT} \tag{6}$$

An implicit assumption of the Larson-Miller analysis is that the activation energy Q is a function of stress so that the Larson-Miller parameter

$$P_{LM} = T\ (\log t_f + C) \times 10^{-3} \tag{7}$$

is a monotonic function of stress alone. Figure 5 shows a typical example of a Larson-Miller plot: typically C takes values in the range 15 to 25. This is a pragmatic approach that allows mechanisms fields to be traversed by virtue of the variation in Q.

An alternative parametric representation due to Sherby and Dorn[11] depends on Q being constant and so is likely to be appropriate when a single mechanism dominates. From Equation (6) in this case:

$$\log t_f - \frac{Q}{RT} = \log A + \log f(\sigma) \tag{8}$$

$$= g(\sigma), \text{ a function of stress alone}$$

Thus the Sherby-Dorn parameter P_{SD}, where

$$P_{SD} = \log t_t - \frac{b}{T} \tag{9}$$

can also give a common curve of data at different stresses and temperatures when plotted as a function of stress.

Design against creep has fallen into two quite different categories. In design codes the material must exceed some measure of creep performance which might be expressed as the stress to give a specified strain (or rupture) in a certain time (say, 1000 h rupture stress). Analytical design methods, on the other hand, attempt to calculate the strains in a component from equations describing the creep behaviour. Both are used quite extensively and operate on a database of constant load and temperature creep curves. Clearly, the assumption of a steady-state creep behaviour for complex alloys leads to overestimates in performance. The level of sophistication of analytical design, enabled by development in finite element modelling (FEM), for example, depends on the ability to accurately represent the full shape of creep curves under varying stress and temperature conditions. This is one of the most active areas of research on creep of engineering materials.

There have been many attempts to represent the shapes of creep curves that have met with varying degrees of success. Examples of the equations describing the primary and steady state creep of simple metals have been given by among others, Andrade,[12] Garafalo[13] and Webster et al.[14] For example, ignoring the elastic strains:

$$\epsilon = \alpha t^{1/3} + \dot{\epsilon}_{min} t \tag{10}$$

or
$$\epsilon = \epsilon_p \left[1 - \exp(-c\dot{\epsilon}_{ss}t) \right] + \dot{\epsilon}_{ss}t \tag{11}$$

have been widely used. Equation 11 is the direct consequence of a recovery-controlled creep model.[23] It envisages the dislocation density that evolves during creep giving rise to an internal stress σ_i that resists creep

$$\dot{\epsilon} = A (\sigma - \sigma_i)^n \tag{12}$$

However, the rate of change of σ_i has a positive contribution due to new dislocations being created with increasing strain, and a negative component associated with dislocations being destroyed by recovery which increases with increasing σ_i

$$\dot{\sigma}_i = H\dot{\epsilon} - R\sigma_i \tag{13}$$

The constants H and R are the hardening and recovery coefficients respectively. Ion et al[5] show that Equations 12 and 13 integrate to give Equation 11.

The current challenge is to provide reliable descriptions of the full strain-time evolution of complex alloys where tertiary creep dominates. There have been several approaches to this problem based on both empirical and physical models

(i) As early as 1953, Graham and Walles[16] proposed a power-law empirical equation describing the full shape of a creep curve - primary, steady state and tertiary.

$$\epsilon = at^m + bt^n + ct^P \qquad (14)$$

They developed a simple graphical procedure for identifying the parameters a, b and c of the equation from a measured creep curve; the exponents were taken to have values $m=\frac{1}{3}$, n=1 and p=3 from experience in analysing several materials.

(ii) More recently Evans and Wilshire[17] have proposed an alternative equation that is capable of describing a wide range of creep curve shapes.

$$\epsilon = \theta_1 \left[1 - \exp(-\theta_2 t)\right] + \theta_3 \left[\exp(\theta_4 t) - 1\right] \qquad (15)$$

where θ_1, θ_2, θ_3, θ_4 are constants for a given creep curve.

This approach, known as the θ-projection method involves the use of sophisticated computational codes to evaluate the θ-parameters and to correlate their variations with stress and temperature. An example of the fit of Equation 15 to a typical creep curve of IN100, a nickel-base superalloy, is given in Figure 6.

Equations which express creep strain as functions of stress, temperature and time, such as Equations 14 and 15, can provide an accurate description of constant load and constant stress creep curves. However, if calculations involving changing stresses or temperatures are required then arbitrary rules defining the transitions between creep curves are required as indicated in Figure 7. In principle it is possible to proceed between creep curves by an infinite number of paths between the extremes of time hardening (path A) and strain hardening (path B). However, for superalloys the strain hardening criterion is appropriate.

(iii) In a joint NPL/Cambridge University project[15,18] an alternative approach designated CRISPEN and operating on a personal computer has been developed in which equations are used to describe the strain rate, rather than strain, in terms of state variables which have clear physical significance in terms of the known mechanisms of deformation and fracture. This approach is based on the concepts of continuum damage mechanics developed by Kachanov,[19] Hayhurst and Leckie[20] and others. However, it has the novel feature, in relation to CDM, that the specific equations used are informed by the known damage mechanisms such as those recently reviewed by Ashby and Dyson.[21]

The creep curve is represented by a series of coupled differential equations which describe the evolution of strain and of state variables S_1, S_2, ... (which may be, for example, particle sizes, dislocation densities, cavitation densities etc. depending on the controlling mechanisms): In general,

$$\left.\begin{aligned}
\dot{\epsilon} &= f(\sigma, T, S_1, S_2, \dots) \\
\dot{S}_1 &= f(\sigma, T, S_1, S_2, \dots) \\
\dot{S}_2 &= f(\sigma, T, S_1, S_2, \dots)
\end{aligned}\right\} \qquad (16)$$

For the case of nickel-base superalloys and other engineering alloys the following specific equations are found to apply over a wide range of conditions.

$$\left.\begin{aligned}
\dot{\epsilon} &= \dot{\epsilon}_{ref}(1-S_1)(1+S_2) \\
\dot{S}_1 &= H\dot{\epsilon}_{ref}(1-S_2) - RS_1 \\
\dot{S}_2 &= C\dot{\epsilon}
\end{aligned}\right\} \qquad (17)$$

The creep curve is determined by numerical integration of Equation set 17 using standard computational techniques which can easily cope with complex stress/temperature histories. As with the θ-projection approach, an individual creep curve is completely described by four parameters ($\dot{\epsilon}_{ref}$, H, R, C) which can vary with stress and temperature. Typical examples are shown in Figure 8 for three different superalloys. However, because the evolution laws for the state variables are explicitly incorporated in the model, it is quite straight forward to calculate creep curves for quite complex stress and temperature profiles as shown in Figure 9, provided, of course, that additional fatigue damage is not introduced when the conditions change.

Winstone[22] has recently compared these three approaches by using them to analyse the same database of creep curves for the single crystal superalloy SRR99 with [100] tensile orientation. Each creep curve was analysed to give the appropriate sets of model parameters which were then expressed as functions of stress and temperature. Using these analytical expressions for the parameters, the lives for each test condition were calculated and compared with the measured values. These comparisons are shown in Figure 10 indicating that there is little to choose between the various models for representing constant stress/constant load creep data. However, the more demanding test is the prediction of strain accumulation during more complex loading conditions (cyclic, multiaxial, simultaneous corrosion[40]).

As materials became increasingly complex and anisotropic it is important that the computational approaches are sufficiently flexible to accommodate the characteristics of these materials. Two examples can be given of the modification of CRISPEN and the CDM formalism to account for creep in single crystal superalloys and in metal matrix composites.

Ghosh and McLean[23] have reformulated Equations 17 in terms of shear stresses and strains on the operative slip systems to allow creep curves for arbitrary crystal orientations in single crystal superalloys to be calculated. In these materials creep strain can lead to large crystal rotations which can also be calculated. Indeed, experimental measurement of the rotations should provide a discriminatory test of the validity of the model. Figure 11 shows a sample calculation for SRR99 when both octahedral and cube slip operate.

McLean[24] and Goto and McLean[25] have formulated equations representing the time dependent deformation of metal matrix composites which are composed of phases having quite different characteristics. Thus, the high melting point ceramic fibres will deform elastically while the low melting point matrix will be subject to creep deformation. The CDM formalism allows the effects of a variety of types of damage to be introduced and their effects to be assessed. Thus fibre aspect ratio, fibre alignment and fibre/matrix interface strength have all been considered. Figure 12 shows the result of sample calculation showing the influence of fibre/matrix interface strength on the creep curves of a metal-matrix composite.

4   Fatigue

The development of models of fatigue of high temperature materials follows similar trends to those described for creep. For simpler design codes a knowledge of the endurance limits of the materials subject to various stress or strain controlled cycles is adequate. Here the number of cycles to failure N_f has a clear parallel to the rupture life in creep. There is no attempt to understand or represent the behaviour leading to failure and a series of empirical rules, or models, have been developed and extensively used for purposes of alloy development or engineering design.

However, as design procedures have developed, much more detailed descriptions of the fatigue behaviour are required which involve, for example a knowledge of the crack initiation and growth behaviour and of the cyclic hardening or softening associated with arbitrary stress and/or temperature cycles which may be in or out of phase. Indeed, the problems of accounting for the combined effects of creep and fatigue (creep/fatigue interactions) through a unified model must be the target for future design procedures using numerical techniques to simulate the likely service performance.

Endurance behaviour

Tomkins[26] and Chaboche[27] have reviewed the basic concepts of fatigue of high temperature materials. In a strain controlled fatigue test, the endurance behaviour is presented as in Figure 13 as the number of cycles to failure as a function of the applied strain range. (Similar plots for load controlled tests are expressed in terms of the load range.) The region up to about 10^4 cycles is usually referred to as high strain or low cycle fatigue (LCF) and is of particular interest in relation to operational cycles of high temperature plant. The low stress/high cycle fatigue (HCF) behaviour will not be considered in this discussion.

In any fatigue cycle the total strain range $\Delta\epsilon$ has elastic and plastic components, as is shown schematically in Figure 14. The plastic component is particularly important in LCF, while with local plastic strain restricted to stress concentrators, HCF is essentially an elastic-strain cycling phenomenon. At the extremes the endurance behaviour can be represented by power-law equations as proposed by Coffin[28] and Manson[29] The cumulative effect is obtained by summing the effects of elastic and plastic strain cycling ($\Delta\epsilon_{el}$ and $\Delta\epsilon_{pl}$ respectively).

$$\Delta\epsilon_{tot} = \Delta\epsilon_{el} + \Delta\epsilon_{pl} \tag{18}$$
$$= C_1 N_f^{-\alpha} + C_2 N_f^{-\beta}$$

137

where N_f is the number of cycles to fracture. C_1 and C_2 can be broadly associated with the tensile fracture strength and ductility of the material; α and β are adjustable parameters.

At elevated temperatures the influence of creep can make a significant reduction in the fatigue resistance. The concept of strain range partitioning was introduced to separate the effects of creep and fatigue damage and to this end four types of strain elements are introduced because of their differing effects on the generation of creep damage - $\Delta\epsilon_{cc}$, $\Delta\epsilon_{pp}$, $\Delta\epsilon_{cp}$, $\Delta\epsilon_{pc}$ where the sub-scripts p and c represent time-independent plastic and creep strain respectively. Four separate failure equations are identified with each of these strain modes

$$N_{ij} = C_{ij} \, \Delta\epsilon_{pt}^{-\delta_{ij}} \qquad (19)$$

where i,j = p or c

In a complex cycle these are combined to give an overall endurance in terms of an empirical interaction rule such as:

$$\frac{1}{N_f} = \frac{1}{\Delta\epsilon_{pt}} \left[\frac{\Delta\epsilon_{pp}}{N_{pp}} + \frac{\Delta\epsilon_{cc}}{N_{cc}} + \frac{\Delta\epsilon_{pc}}{N_{pc}} + \frac{\Delta\epsilon_{cp}}{N_{cp}} \right] \qquad (20)$$

Damage accumulation

Where there are significant hold times at high temperatures the effects of creep and fatigue must be combined and this is often achieved by use of a life fraction rule. If N_f is the fatigue life in the absence of creep and t_f is the rupture life in the absence of fatigue then failure is taken to occur when

$$\frac{N}{N_f} + \frac{t}{t_f} = 1 \qquad (21)$$

These parametric approaches are now incorporated in different design codes. However, the uncertainties in accounting for random fatigue cycles and creep-fatigue interactions, for example, require the introduction of large safety factors which militate against efficient design.

Continuum damage mechanics is currently being applied by Chaboche and Lemaitre[30] from the perspective of the

138

fatigue problem to provide a rational basis for accounting for creep fatigue interactions. The approach is to define two empirical damage evolution laws from pure creep and from pure fatigue tests to account for the differing nature of highly localized fatigue and non-dispersed creep damage.

a) For fatigue

$$dD_F = f(\sigma_m, \bar{\sigma}, T, D_F)dN \qquad (22)$$

where σ_m and $\bar{\sigma}$ are the maximum and mean stresses.

b) For creep

$$dD_C = g(\sigma, T, D_C)dt \qquad (23)$$

The calculations for mixed fatigue and creep depend on two assumptions

(i) D_F and D_C are additive so that

$$dD_F = f(\sigma_m, \bar{\sigma}, T, D_F + D_C)dN \qquad (24)$$

$$dD_C = g(\sigma, T, D_F + D_C)dt \qquad (25)$$

(ii) Failure occurs when $D_F + D_C = 1$

Numerical integration of Equations 24 and 25 gives the number of cycles to crack initiation. Comparison of such calculations with data for IN 100 are shown in Figure 15.

A more physics-orientated approach to combining the effects of combined creep and fatigue damage has been proposed by Majundar and Maiya.[31] Here explicit expressions are written describing the growth of fatigue cracks a and the increase in the density of creep cavities c that are compatible with detailed experimental studies of fatigue cracks initiation and growth. Here, the damage synergism is taken to be one-sided; creep cavities are believed to influence fatigue crack growth but fatigue damage is not considered to affect creep damage.

For fatigue:

$$\frac{da}{dN} = A\left(1 + \alpha \log \frac{c}{c_0}\right) |\epsilon_p|^m |\dot{\epsilon}_p|^k a \qquad (26)$$

139

For creep:

$$\frac{dc}{dt} = \gamma \, B \, |\epsilon_p|^m \, |\dot{\epsilon}_p|^{k_c} \, c \qquad (27)$$

A, α, β, B, m and k are all constants.

As with creep modelling it is necessary to maintain a degree of empiricism in order to keep the equations sufficiently simple to be useful and to be capable of being extended to deal with factors such as multiaxiality and the effects of environment on crack growth characteristics. However, further developments are likely to result from a clearer understanding of the synergies between various types of damage that will result from detailed mechanistic studies and the associated micro-mechanic modelling. In this context the work of Hutchinson, Rice and Riedel[32] may prove to be of particular importance in establishing models of macroscopic crack growth in the presence of developing creep cavities.

Little information is available about thermal mechanical fatigue because of the practical difficulties in carrying out the tests.[33] However, this is recognized as a topic of considerable importance and models must be fully capable of accounting for variable stress and temperatures that are in or out of phase. Figure 16 shows some data for the nickel-based superalloy IN738LC. The results fall at the lower end of the isothermal scatterband with the in-phase results being the most damaging.

5 Discussion

The separation of creep and fatigue models is a matter of convenience for the scientists working in these fields. The engineering requirement is quite clearly for a unified model which can be applied to the engineering problems that are likely to arise in service. It is likely to be a long time before this objective is achieved and even longer before design engineers will be persuaded to use the techniques. The work at ONERA has made significant advances on this topic from an empirical viewpoint. A more mechanistic approach, relevant to simple metals is being taken by Miller[34] who has developed computational techniques for predicting high temperature fatigue cycles for materials that exhibit extensive primary creep.

Ghosh and McLean[35] are currently extending the continuum damage description of high temperature deformation in engineering alloys, described in Section 3. It envisages the mechanisms of deformation as being similar in creep, tensile, stress relaxation and cyclic tests but that the controlling parameter in each test establishes a different boundary condition for the calculation of strains. Equation set 17 is generalised to:

$$\dot{\sigma} = E\,(\,\dot{\epsilon} - \dot{\epsilon}_{cr}\,)$$
$$\dot{\epsilon}_{cr} = \dot{\epsilon}_{ref}\,(1 - S_1)\,(1 + S_2)$$
$$\dot{S}_1 = H\,\dot{\epsilon}_{ref}\,(1 - S_1) - RS_1 \qquad\qquad (28)$$
$$\dot{S}_2 = C\,\dot{\epsilon}_{cr}$$

where E is Youngs modulus.

In a creep test $\dot{\sigma} = 0$, and Equation set 28 reduces to Equation set 17, as it should. Other types of tests can be simulated by setting the appropriate conditions:

constant strain rate   — $\dot{\epsilon}$ = constant

stress relaxation   — $\dot{\epsilon}$ = 0

strain controlled fatigue   — $\epsilon(t)$

load controlled fatigue   — $\sigma(t)$

Preliminary examples of simulations of such tests using model parameters determined by the analysis of creep data for the nickel-base superalloy IN738LC are shown in Figure 17. Figure 17a predicts the establishment of a peak stress, at about 1% strain, that decreases with decreasing strain rate; thereafter the stress progressively falls. This type of behaviour has been reported by Gibbons and Hopkins[41] for model γ-γ' alloys. For the case of strain-controlled low cycle fatigue (Figure 17b) the model can predict the observed shape of creep curve and a progressive cyclic softening behaviour that leads to a decrease in the stress range with increasing number of cycles. These figures are intended to indicate the directions in which the models may evolve rather than suggesting that a formal model is already available to describe creep-fatigue behaviour. Much more work is required to establish and validate the form and parameters of the model. However, the availability of

141

increasing computer power will accelerate development in such modelling and will create a demand by designers for more detailed descriptions of the material behaviour in potential service conditions.

6   REFERENCES

1.   H.J. Frost and M.F. Ashby, "Deformation-Mechanism Maps", Permagon Press, Oxford 1982.
2.   D.McLean, Reports of Progress in Physics 29, 1 (1966).
3.   R. Lagneborg, Int. Metall. Rev. 17, 130 (1972).
4.   B.F. Dyson and M. McLean, Acta Metall. 31, 17 (1983).
5.   T.S. Lundy and J.F. Murdock, J. Appl. Phys. 33, 1671 (1962).
6.   M. Maldini, A. Barbosa, B.F. Dyson and M. McLean, "Analysis of creep data from constant load and from step- and cyclic-load tests", NPL report DMA(A)116, January 1987.
7.   O.D. Sherby and P.M. Burke, Progress in Materials Science 13, 325 (1967).
8.   H. Burt, J.P. Dennison and B. Wilshire, Metal Science 13, 295 (1979).
9.   F.C. Monkman and N.J. Grant, Proc. ASTM 56, 593 (1956).
10. F.R. Larson and J. Miller, Trans ASME 74, 765 (1952).
11. R.L. Orr, O.D. Sherby and J.E. Dorn, Trans ASM 46, 113 (1954).
12. E.N. da C. Andrade, Proc. Roy. Soc. A84, (1910).
13. F. Garafalo, "Fundamentals of Creep and Creep-Rupture in Metals," Macmillan, New York 1965.
14. G. A. Webster, A.P.D. Cox and J. E. Dorn, Metal Sci J. 3, 221 (1969).
15. J.C. Ion, A. Barbosa, M.F. Ashby, B. F. Dyson and M. McLean, "The modelling of creep for engineering design I" NPL report DMA(A)115, April 1986.
16. A. Graham and K.F.A. Walles, J. Iron Steel Inst. 179, 105 (1955).
17. R.W. Evans and B. Wilshire, "Creep of Metals and Alloys", Institute of Metals, London 1985.
18. A. Barbosa, N.G. Taylor, M.F. Ashby, B. F. Dyson and M. McLean in "Superalloys 1988" edited by D. N. Duhl et al. pp 683-692, The Metallurgical Society Inc. Warrendale, Pa. 1988.
19. L.M. Kachanov, Izv. Akad. Nauk. SSR No.8, 16 (1958).

20. F.A. Leckie and D.R. Hayhurst, Acta Metall. 25, 1059 (1977).
21. M.F. Ashby and B.F. Dyson, Proceedings of 6th International Conference on Fracture (ICF6), New Delhi, Permagon, p3 (1984).
22. M.R. Winstone, MTU symposium on "Single crystals for turbine blades", Munich, June 1989.
23. R.N. Ghosh and M. McLean, Scripta Metall. 23, 1301 (1989).
24. M.McLean, in "Materials and Engineering Design - the next decade", edited by B.F. Dyson and D.R. Hayhurst, The Institute of Metals, London 1989.
25. S. Goto and M. McLean, Scripta Metall. to be published.
26. B. Tomkins, in "Creep and Fatigue in High Temperature Alloys" edited by J. Bressers, Applied Science 1981.
27. J.L. Chaboche, High Temperature Technology 5, 59 (1987).
28. L.F. Coffin in "Proceedings of International Conference on Fracture - 2" p. 643 Chapman and Hall, London (1969).
29. S.S. Manson, ASTM STP520, p744 (1972).
30. J. Lemaitre and J.L. Chaboche, "Mechanics of Visco-Plastic Media and Bodies," edited by J. Hult, Springer Verlag, Berlin, pp.291-301 (1975).
31. S. Majundar and P.S. Maiya, J. Eng. Mater. Tech. 102, 159 (1980).
32. H. Riedel "Fracture at High Temperatures," Springer Verlag, Berlin 1987.
33. W.J. Plumbridge and E.G. Ellison, Mater. Sci. Tech. 3, 706 (1987).
34. A.K. Miller, in "Unified Constitutive Equations for Creep and Plasticity," ed. by A.K. Miller, pp 139-219, Applied Science 1987.
35. R.N. Ghosh and M. McLean, to be published.
36. J.B. Conway, "Numerical Methods for Creep and Rupture Analysis", Gordon and Breach, New York 1967.
37. R.W. Evans and B. Wilshire, Materials Science and Technology 3, 701 (1987).
38. G.E. Leese and R.G. Bell, paper 15, Proc. of AGARD Conference No 393, 1985.
39. M.R. Winstone, MTU Symposium on "Single Crystals for Turbine Blades," Munich, June 1989.
40. B.F.Dyson and S.Osgerby in "Materials and Engineering Design - the Next Decade", edited by B.F.Dyson and D.R.Hayhurst, The Institute of Metals, London (1989).
41. T.B.Gibbons and B.E.Hopkins, Metal Science 8, 203 (1974)

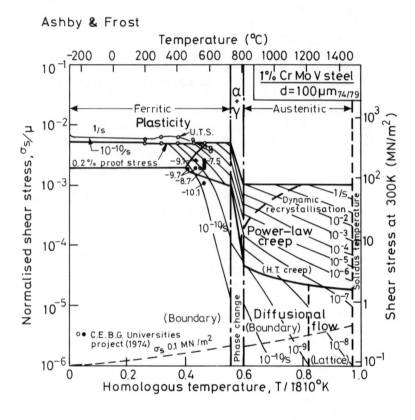

Figure 1

Deformation mechanism map for 1% Cr-Mo-V steel of grain size 100μm.

144

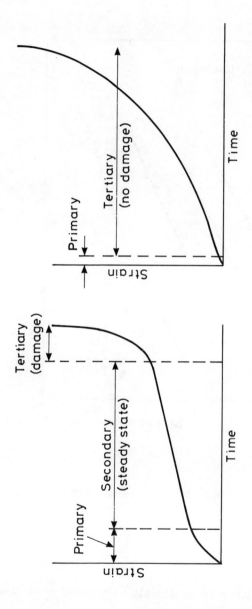

Figure 2

Schematic diagrams showing the general features of creep curves for (a) simple metals and (b) engineering alloys

145

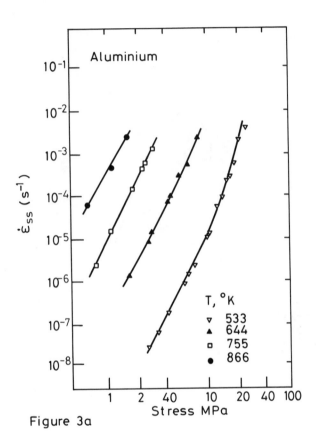

Figure 3a

Figure 3

Steady state or minimum creep rate as a function of stress
at various temperatures:
a) Pure polycrystalline aluminium[5]
b) Directionally solidified IN738LC.[6]

146

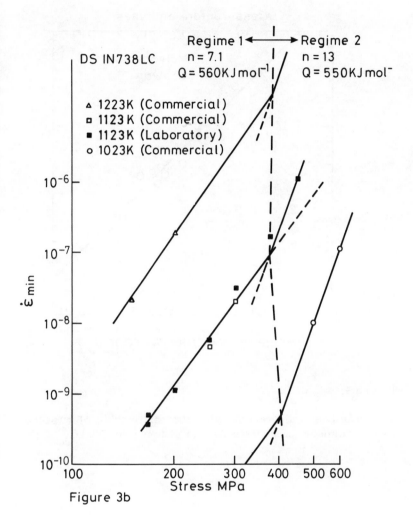

Figure 3b

147

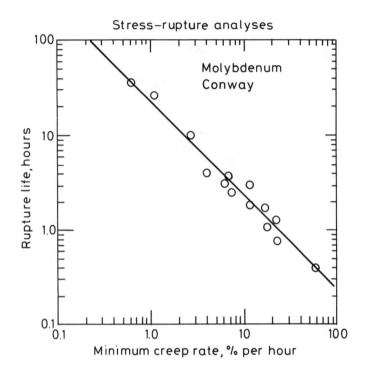

Figure 4

Monkman-Grant diagram showing the relationship of rupture life to minimum creep rate for molybdenum at 2000°C.[36]

148

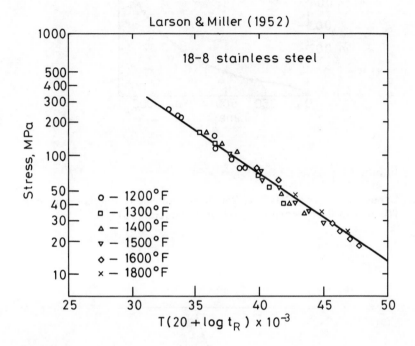

Figure 5

Larson-Miller diagram showing the variation of the compound parameter $T(20 + \log t_f) \times 10^{-3}$ as a function of stress for 18-8 stainless steel.[36]

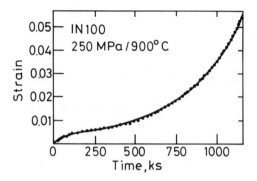

Figure 6

Creep curve for IN100 at 250 MPa and 900°C together with a fit of the θ-projection equation (Equation 15).[37]

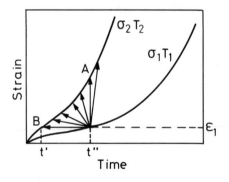

Figure 7

Schematic diagram showing various rules for defining the strain on changing stress and/or temperature during a creep test.[37]

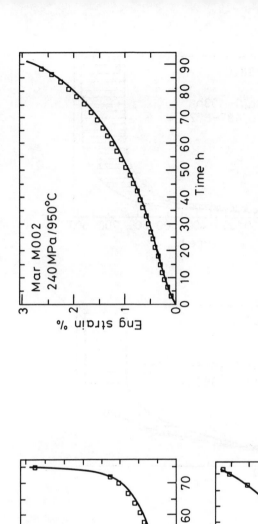

Figure 8

Examples of the agreement of the experimental data with the CRISPEN representation of creep deformation (Equation set 17)

a) IN738LC. directionally solidified, 375 MPa/850°C
b) Mar M002, conventionally cast, 240 MPa/950°C
c) SRR99, single crystal, 600 MPa/900°C.

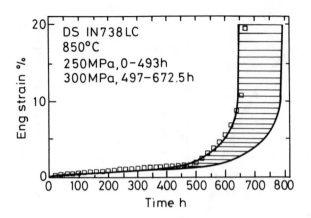

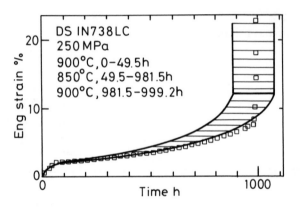

Figure 9

As figure 8, but for changing stress and temperature
a) σ_1=250 MPa, 0-497h; σ_2=300 MPa, 497-672.5h, T=850°C
b) σ=250 MPa
     T_1=900°C, 0-49.5h; T_2=850°C, 49.5-981.5h; T_3=900°C, 981.5-999.2h.

152

Predictions of rupture life for (100) oriented SRR99

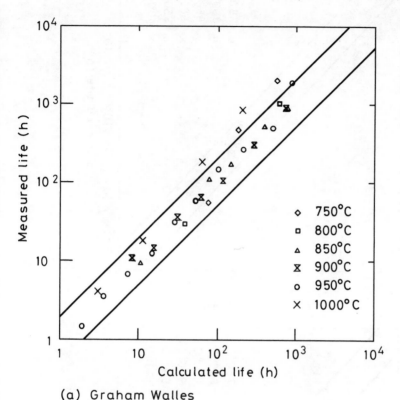

(a) Graham Walles

Figure 10

Comparison of measured and calculated rupture lives for a
creep database of SRR99 analysed by different models of
creep curve shape (after Winstone[39])
a)   Graham-Walles
b)   θ-projection   } (shown on p.154)
c)   CRISPEN.

153

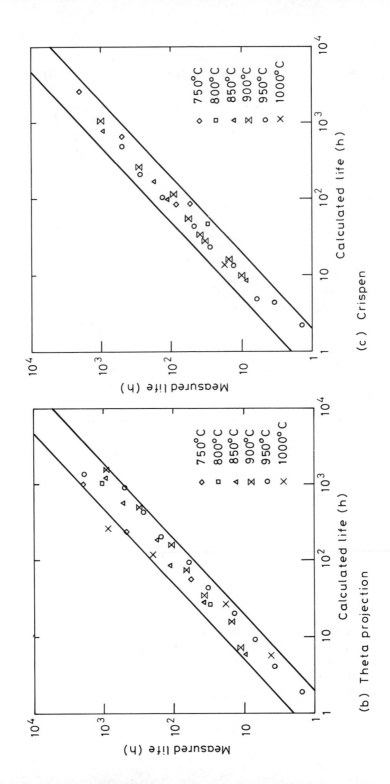

(b) Theta projection

(c) Crispen

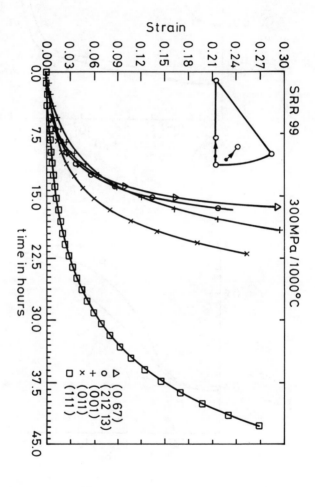

Figure 11

Illustrative calculation showing the likely effect of combined {111}<1l0> and {100}<011> slip for SRR99 at 1000°C and 300 MPa assuming the shear strain rate on the cube plane is 0.25 of that on the octahedral plane.[23]

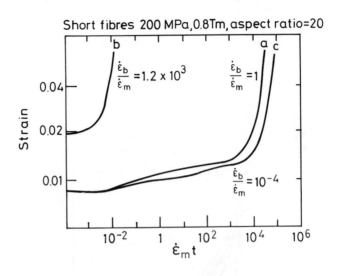

Short fibres 200 MPa, 0.8Tm, aspect ratio=20

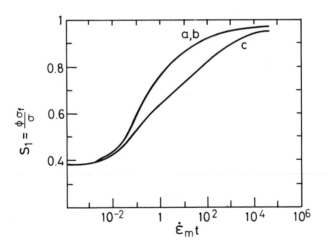

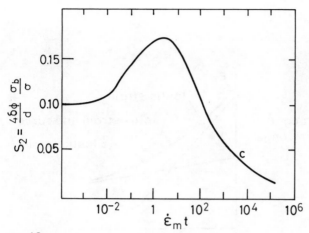

Figure 12

Predicted creep curves and stresses on fibre and
fibre/matrix boundary for an aluminium/20% SiC metal matrix
composite with fibre aspect ratio of 20. The effects of
weak interfaces and of a work hardened boundary zone are
shown together with the curves for the ideal composite.
Stress= 200 MPa, Temperature = $0.8 \, T_m$.[25]

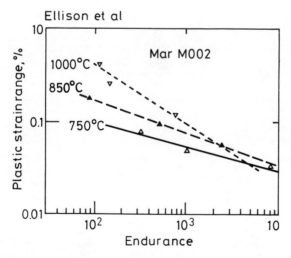

Figure 13

Effect of temperature on the endurance as a function of
plastic strain range in Mar M002.[33]

157

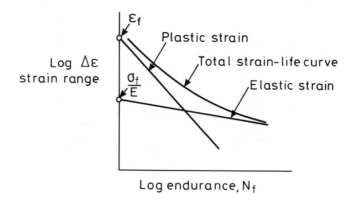

Figure 14

Schematic illustration of fatigue life as a function of elastic, plastic and total strain amplitude.

158

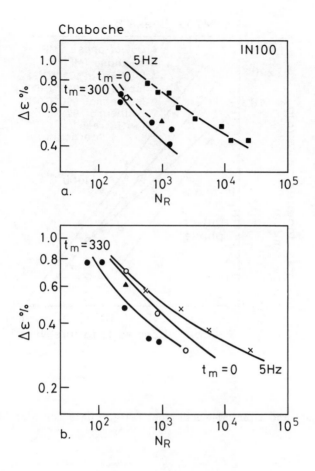

Figure 15

Prediction of the results of strain controlled cyclic tests
with differing hold times[27] on IN100 by the non-linear
continuous damage model.

159

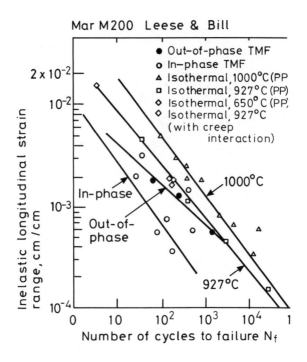

Figure 16

Isothermal and thermal-mechanical fatigue data for Mar M-200.[38]

Figure 17 (shown on pp.161-162)

Stress-strain curves for (a) a constant strain-rate tensile test and (b) strain controlled cycling calculated from equation set 28 using parameters established from a creep database IN738LC at 850°C.

160

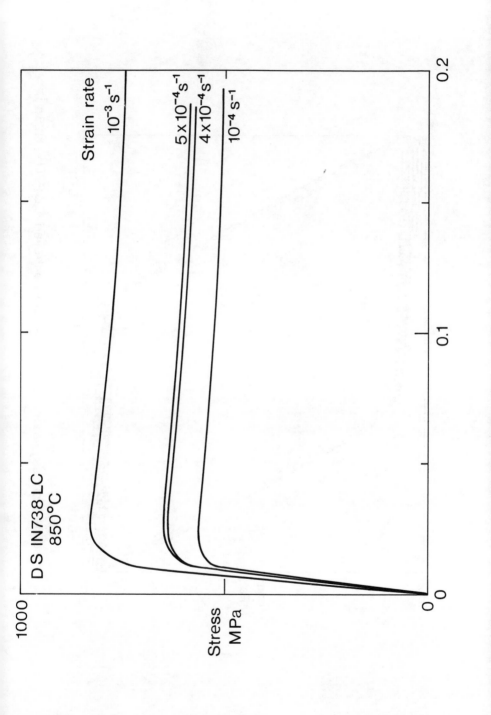

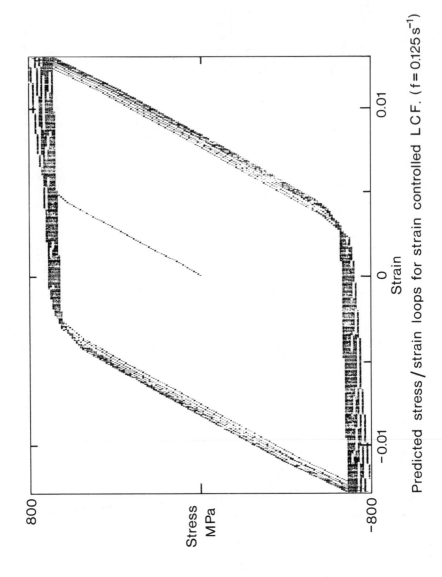

Predicted stress / strain loops for strain controlled L.C.F. ($f = 0.125 \, s^{-1}$)

6:Process modelling

O L TOWERS

Rolls Royce Plc, Derby

1. INTRODUCTION

The cost of developing a new jet engine is of the order of $1bn.[1] At the same time the lead times between conceptual design and production run into many years. This means that the risk has to be taken of a large outflow of funds some years before the product reaches the market. These commitments are such that both airframe and jet engine manufacturers have at times been in severe financial difficulties in part due to the problems experienced with controlling jet engine development lead times and costs. This was particularly the case with respect to the development of the jet engines for the new wide-bodied aircraft in the late 1960's and early 1970's.[1]

It is essential to control and reduce the above development lead times and costs. One important element of this is to reduce the extent of trial manufacture and testing before an acceptable design is arrived at. Such iterative trials waste materials and labour and can interrupt normal production, whilst adding to the development lead time. The way to avoid such problems is to 'test' the design options on a computer. In this way, provided the computer can perform the appropriate analyses faster than the trial manufacture and testing route, significant reductions in lead time and cost can be made. This approach has been used to design components for service conditions using (largely) linear elastic stress analyses and thermal analyses for some years. Indeed some of the pioneering work on finite element based linear elastic stress analysis was performed by airframe and jet engine manufacturers in the mid to late 1950's and early 1960's.

The development of numerical analysis codes to solve non-linear transient problems, as described in the previous paper,[2] together with the increasing power of computers now means that is is possible to simulate the behaviour of materials in manufacturing processes, ie to perform 'process modelling'. This, together with the acquisition of basic data on the behaviour of materials at the conditions relevant to the manufacturing process as well as characterisation of the process itself, means that trial and error design of the manufacturing process can be performed on the computer rather than by making parts and testing them. The importance of these developments should not be underestimated. For instance in forging a turbine disc for a jet engine in a new turbine disc alloy, process modelling was used to reduce the lead time from two years to four months largely because the manufacturing route developed on the computer resulted in the first disc having the required microstructure, ie being 'right-first-time'.[3]

In this paper the requirements of a process model are reviewed with respect to how to reduce lead times through close integration between design and manufacture. The particular needs which a process model has to meet are then covered for the important basic manufacturing processes of casting, forging and heat treatment. Within this the types of analysis package required are discussed, leading to a review of the data which are required to perform the analyses for these processes. Two particular examples involving the modelling of casting and forging are described. These exercises lead into a review of the current limitations of process models, together with a view of the future developments predicted in the process modelling field.

The paper is generally oriented towards modelling the manufacture of high temperature components for jet engines, although the principles described apply equally to the many other industries where forged, cast and heat treated parts are made and used. For those wishing to obtain more in-depth knowledge on process modelling a reading list of relevant books or major publications is provided in the appendix.

2. WHAT IS REQUIRED FROM A PROCESS MODEL ?

To minimise lead times and reduce costs, the ultimate objective must be for the first component manufactured to meet the required specification, ie to be 'right-first-time'. To achieve this not only must the manufacturing process produce a component which is within tolerance, sufficiently free of defects to meet the specification and with a microstructure which has the required properties, but also the design must be such that it is possible for the component to be made in the first place.

To succeed in 'right-first-time' manufacture requires close integration between the design of the component, to meet the needs of service conditions, and the design of the manufacturing process, to be compatible with the production facilities. The traditional iterations involved in developing a manufacturing process route for a component, with consequence penalties in lead times and costs, can be in part attributed to the barrier which has tended to exist between

165

design functions and production areas (fig 1). These barriers have meant that designers have not taken due consideration of the problems associated with manufacture, and have also meant that production areas have tended to take on work without the knowledge of how to produce the component.

The importance of process modelling is that it both provides a means to check that proposed designs are 'makeable' and it also provides the means for manufacturing process options to be tested on the computer to come up with the optimum route. This means that process modelling lies on the critical interface between component design and process design and is the means of ensuring that these two activities are integrated.

It is worth noting that a process model requires the person using it to make decisions on the particular conditions being simulated. Thus the component geometry and its 'attributes' (for instance material constituents) must be specified, together with relevant materials' property data and the boundary conditions relevant to the manufacturing process. These data are then used as input to the process model, as shown schematically in fig 2. The process model cannot itself decide on the geometry to be simulated and does not determine what the process parameters should be (eg forging temperature, forging speed, etc). This means that the process options simulated are by definition limited by the knowledge and imagination of the user of the model.

In the long-run the use of expert systems can be envisaged where a decision is made by a computer program as to the relevant process parameters to model. Indeed the ideal would be for the system to itself select the optimum process parameters. In the short to medium term, however, process models are likely to be used for 'what if' type calculations, in an analogous manner to the use of spreadsheets for 'what if' calculations in financial applications. Although thus limited in scope, the potential for the use of process models in this 'what if' mode is considerable. Certainly, the use of process modelling even in this restricted mode of operation is in its infancy and the benefits are only starting to be realised. Further, as computers become faster, the disadvantages of performing calculations in this 'forward' mode will be minimised and the ability of the models to reduce lead times will increase.

3.   TYPES OF MODEL REQUIRED FOR DIFFERENT MANUFACTURING PROCESSES

Potentially, the lead time and cost benefits accruing from rapid computer-based simulations can be realised for any manufacturing process. There are two important general factors which dictate the usefulness of a model in practice. Firstly, the importance of the manufacturing process, and particularly the lead times required for process route development, to a given business and secondly the complexity of simulating the process with sufficient accuracy to provide a predictive model.

In the context of high temperature components in a jet engine, basic manufacturing processes such as casting, forging, heat treatment, and welding are still used to produce the majority of components despite the obvious attractions of composites, ceramics and other alternatives.[4] Turbine discs are forged from superalloy or titanium alloy billets. Turbine blades are investment cast in superalloys, increasingly with directionally solidified microstructures. Compressor blades, which themselves can reach relatively high temperatures (approximately 500-600°C), are forged in a variety of alloys including superalloys and titanium alloys. The discs are joined together in drum assemblies using electron beam welding. Furthermore the majority of these parts are heat treated to achieve the required properties.

The features of particular interest in the manufacturing processes of casting, forging and heat treatment are discussed in turn below, focusing particularly on the consequent requirements for a predictive process model.

3.1 Casting

As already mentioned turbine blades are produced by investment casting, with an increasing proportion being produced with directionally solidified or single crystal microstructures (fig 3). For a cast blade to be acceptable it must be 'sound' (ie free of macroscopic porosity) and free of metallurgical defects which may be deleterious to the component's service performance. Furthermore, the blade must be within certain dimensional tolerances.

Jet Engine casings are also produced in superalloys and titanium alloys. These have equiaxed microstructures. Again the objective is to produce sound castings, without unacceptable metallurgical defects and which are within dimensional tolerances.

Castings of course are used in many other industrial applications involving high temperatures. Examples include steam turbines, where relatively large steel castings are used for a variety of purposes, and the automotive industry where the turbines for turbochargers are often investment castings in a nickel-base alloy.

Clearly, if a process model is going to ensure that the first casting is going to be right-first-time then it needs to be capable of predicting:

168

a) if the casting will be sound,

b) if metallurgical defects are going to occur,

c) the microstructures developed, and

d) dimensional distortion.

3.1.1 MOULD FILLING

The starting point of a rigorous model of casting is the simulation of mould filling. Indeed certain defects in castings, eg 'cold shuts' and 'misruns', occur during the filling operation. Furthermore, the solidification sequence of the casting, which itself will be related to the formation of metallurgical defects, may begin due to cooling during mould filling (particularly for complex, thin section castings where the mould temperature is below the alloy solidus temperature).

To simulate mould filling requires a fluid flow package capable of:

a) dealing with free surfaces,

b) three dimensional analyses (the vast majority of castings are 3D),

c) handling a liquid with a viscosity which changes rapidly with temperature, and

d) carrying out, or interfacing with, a thermal analysis which incorporates the evolution of latent heat.

State-of-the-art fluid flow packages are available with the above characteristics, although it has to be said that considerable computing power would be needed to perform a mould filling simulation for a casting of the complexity of an intermediate casing in a jet

169

engine (fig 4). A further point is that most fluid flow packages capable of simulating mould filling are based on finite difference techniques,[2] which usually means that the element shapes need to be relatively simple and thus place some limitations on the ability to represent complex geometries.

In discussing the simulation of mould filling for castings, it should be recognised that the control of the filling operation is often quite crude in practice and the filling sequence thus becomes variable. Indeed, random effects, such as splashing, may be present. In such a case process modelling is clearly impractical unless it is used to assess trends. On the other hand there is an increasing tendency in the foundry industry in general to use mould filling practices which ensure that the operation is closely controlled and keep turbulence to a minimum (one example of such a system being given in reference 5). Clearly such processes not only bring benefits in terms of repeatability but also make the modelling of mould filling a practical proposition.

3.1.2 THERMAL ANALYSIS OF CASTING "IN-SITU"

Once the metal is 'in-situ', ie the mould is filled with metal, the casting together with the running system and feeders will solidify (or continue to modify if this started during mould filling). To predict lack of soundness, or metallurgical defects requires a simulation of the heat flow during solidification. As already mentioned, in the general case the starting temperatures for this simulation will depend on heat loss during mould filling. For rapidly filled castings of reasonably large sections or for castings where the mould temperature is initially close to the metal temperature, however, it may be reasonable to ignore the mould filling operation and assume the mould is filled instanteously with metal at the pouring temperature.

The thermal analysis package needed to simulate solidification requires the following characteristics:

   a)   the ability to handle complex 3D shapes,

   b)   representation of the evolution of latent heat,

   c)   material thermal property values which can vary with temperature,

   d)   different material property values for different zones (eg mould and metal),

   e)   the ability to handle radiation heat loss and/or convective heat loss,

   f)   facilities to represent contact resistance between different regions (ideally allowing the resistance to vary during the analysis if mould/metal gaps form), and

   g)   interfaces to a fluid flow package, both to pick up the initial temperatures following the mould filling simulation and also so that thermally driven convection currents occuring once the mould is filled can be represented where appropriate.

In the case of directionally solidified castings (which include single crystal components) the representation of radiation becomes considerably more complicated than for conventional castings because of the progressive withdrawal of the casting through a chamber with a steep temperature gradient.[6]

The results of a thermal analysis of a casting can be used to track isotherms, for example showing the position of the liquidus and solidus temperatures in the casting at a given time. To predict microstructures or the formation of metallurgical defects, however, clearly requires relationships between the predicted temperatures and those features. For instance, considerable work has been done to relate microstructures and certain defects such as porosity to the growth rate of the solidification front (R) and the corresponding temperature gradient (G), e.g. as described by Kurz and Fisher[7] and McLean[8] for microstructure and by Hansen and Sahm[9] for porosity.

Lack of 'soundness' usually means the occurrence of shrinkage defects. These occur as a result of trapping liquid, or 'mushy' zones, in metal which has already solidified. An indication of the position of shrinkage can be obtained by looking for trapped hot spots in the casting. This practice can lead to inaccuracies, however, as it ignores the effect of gravity (the cavities tending to occur at the top of a trapped region). Furthermore, accurate predictions of the position and scale of shrinkage are likely to require the representation of the changing volume and geometry of the casting as shrinkage occurs. Further consideration is given to shrinkage in the casting modelling example in section 5.1.

3.1.3   STRESS ANALYSIS

The final type of analysis which can be beneficial in the modelling of casting is a deformation, and thus stress, analysis of the casting. This will be needed to predict dimensional changes during the process, as well as to predict the occurrence of defects such as hot tears.

172

As for the other analyses, the stress analysis package used needs to deal with complex 3D geometries, multiple materials and material properties which vary with temperature. Indeed the material property options may need to be quite sophisticated, for example:

a)   for invested shells, sintering occurs at the high temperatures which occur in the casting process. This sintering, which causes shrinkage of the shell, is temperature and time dependent.

b)   plasticity in the alloy may need to be represented.

c)   the representation of the alloy's mechanical properties will need to deal with the rapid changes in strength as the alloy solidifies and cools.

3.1.4   OVERVIEW OF CASTING PROCESS
          MODELLING

In summary, the most general process model of casting involves a non-isothermal fluid flow simulation of the mould filling operation, a thermal analysis of solidification once the mould is filled and a stress/deformation analysis. Also a further fluid flow requirement may exist to represent convection driven currents in the liquid metal once the mould is filled. Packages do exist to perform all these types of analysis in 3D. The following points should be noted however:

a)   There is no simple package to do all the analyses at the same time.

b)   The computing time required to do all the above analyses is currently prohibitive for complex components.

c) The basic data produced from the analyses needs to be further interpreted to predict the features of interest.

In practice most modelling of casting to date has concentrated on the thermal simulation of the solidifying metal once a mould is filled, assuming that the metal temperature is uniform at the end of the filling operation. An example of this is given in section 5.1 for a jet engine casing.

There is no doubt that in the future a package will be developed which, when combined with the improvements in computer hardware performance will enable all of the needs to be met for a complete model of casting processes. In the meantime progress needs to be made in gaining an improved understanding of how to model the individual components of the process (mould filling, in-situ solidification, distortion) and also how to interpret the results to address the issues relevant to reducing lead times and costs (ie soundness, metallurgical defects and microstructure).

One approach to the complexities of modelling casting processes is to revert to expert systems, thus quantifying existing expert knowledge, creating a database and rules based on this knowledge and integrating this to establish the best route for making a new casting.[10] This approach, however, will never achieve the benefits needed to dent significantly the lead times and costs built in to the development of manufacturing process routes for new components. The jet engine business, for example, never stays still and existing knowledge, quantified or otherwise, is of limited value if one needs to understand how to cast a component which differs in shape from previous practice, which is cast in a new alloy or which is cast using a different manufacturing process. A predictive tool based on a sound quantitative understanding of casting is essential to tackle these issues. Without it the benefits being sought will never be achieved.

3.2 Forging

In a jet engine both compressor and turbine discs are forged to achieve the required microstructure, and thus properties in the finished component. Other important components which are forged include compressor blades.

Naturally forgings are also used in many other industrial applications involving high temperatures. For example forgings are used in various parts of the high temperature primary cooling circuit for nuclear power plant. Further, forgings are used for steam turbine components, including the turbine blades themselves.

In general, for a process model to predict an acceptable forging route the model needs to be able to predict:

a)   the microstructure,

b)   the formation of defects such as laps,

c)   the compatibility of the forging equipment with the proposed forging route, and

d)   the dimensions of the forged part.

The detailed needs for simulating the forging of axisymmetric shapes (i.e. discs) and the forging of complex 3D parts are discussed below.

3.2.1 FORGING OF AXISYMMETRIC SHAPES (DISCS)

Discs for jet engines are forged 'conventionally', hot die forged or isothermally forged. Conventional forging implies the use of steel dies (which may be at ambient temperature, or preheated to temperatures of up to 300°C). The forging equipment may be a screw press, hydraulic press or a hammer. Hot die forging usually involves using nickel-base dies which are preheated to about 600-700°C. Again the particular equipment used to apply the load can be various. Finally, isothermal forging involves heating the dies to the same temperature as the workpiece. The temperatures involved, which may be up to 1100°C, mean that a molydenum based alloy ('TZM') is used for the dies, which itself means that an inert atmosphere is needed. TZM also introduces handling difficulties due to its brittleness at ambient temperature. Generally, isothermal forging is done in hydraulic presses, which provide the level of control required, as well allowing relatively low rates of deformation.

The simulation of isothermal forging of discs ideally requires the following features in the package:

a) the ability to handle the 2D axisymmetric shapes of the dies and the workpiece,

b) the means of simulating large scale deformation of the workpiece,

c) the representation of the die/workpiece interface with sliding being possible with friction,

d) simulation of the die coming into contact with the workpiece and also

176

the workpiece separating from the die,

e)   the ability to predict stresses, strains and strain rates throughout the workpiece,

f)   the ability to handle material properties where the 'flow' stress is dependent on strain and strain rate in a non-linear manner,

g)   the means to predict microstructure, given its relationship to the variables predicted by the analysis, and

h)   the ability to perform a stress analysis for the dies.

Discs are normally thoroughly inspected after the forging process, which results in the forging shape having a considerable amount of surplus material which is machined off to produce the final component. This means that prediction of the detailed dimensions of the forged part is not normally required. It is worth noting, however, that the moves to nearer net shape forgings, and to inspection of the process rather than the part, may mean that dimensional considerations assume a higher priority in the future.

The majority of the deformation in the forging itself is due to plastic deformation. As such, most packages which are designed specifically to simulate forging assume all the deformation is plastic.[2] Ignoring elastic deformation in the forging is further justified by the above dimensional considerations. The stress analysis for the dies, however, needs to be based on the assumption of small displacement predominantly linear elastic behaviour, because plastic deformations in the die need to be avoided or kept to a minimum (both to avoid die shape changes and to avoid die cracking or wear). The importance of maintaining the die integrity needs to be emphasised for all forgings, but particularly so for isothermal forging where the cost of a set of TZM dies can run into six figures (in sterling).

There are many packages capable of meeting the above needs, and an example of an isothermal disc forging simulation is given in section 5.2. One aspect of isothermal disc forging modelling which it is worth noting, is that because the geometry is two dimensional and no thermal analysis is required, the computer run times are relatively low, at approximately 1 to 3 hours on a state-of-the-art workstation (currently costing perhaps £20,000).

For the simulation of conventional or hot die disc forging the analysis package needs to have the same features as for isothermal forging with the additional requirement for a 'coupled' thermal analysis. This coupling between the thermal and the deformation analyses needs to be sufficiently 'tight' for the heat loss to the dies to be simulated, the heat evolution due to plastic work to be represented and for the strong dependence of the material properties on temperature to be dealt with. The introduction of a thermal analysis also means that the package must provide the facility for representing the workpiece-to-die heat transfer and radiation to the environment. One of the most difficult issues in practice is to know what to assume for the workpiece to die heat transfer coefficients, bearing in mind the complexities of the conditions at the interface (eg die just resting on billet, die applying load, die/workpiece slip occurring with friction, lubricant thinning as forging progresses, etc). A further requirement from the thermal analysis is that it should be capable of representing the heat loss from the billet as it is transferred to and held in the forging

press. Neglecting this can cause errors in
the forging simulation.

Normally the loading history for the forging
operation is represented in the program by a
die velocity (either constant or varying
during the analysis). Where the forging
operation is controlled by servo-hydraulics a
known die velocity history can indeed be
specified and ensured in the forging
operation. In other forging equipment,
however, such controls do not exist or are not
feasible. For example, in hammer forging
energy levels are set for each blow.
Simulating this requires the complex velocity
profile for the hammer to be measured or to be
deduced within the analysis itself (eg by
computing the energy absorbed during forging
and reducing the kinetic energy of the hammer
accordingly). Most available packages would
only be capable of performing such analyses
after modification.

A final point concerning non-isothermal
forging modelling is the computing times
involved. These will obviously depend on the
way in which the thermal analysis is performed
and how tightly coupled it is to the
deformation calculations. As a rough guide,
however, the run times for a non-isothermal
analysis using one particular package are
typically ten times longer than for an
isothermal simulation.

3.2.2   FORGING OF COMPLEX 3D PARTS

Taking compressor blades for jet engines as
examples of complex 3D parts, the principle
differences between the requirements for a
model of compressor blade forging and those
for a model of disc forging are that:

> a)   compressor blade forging modelling
>   requires a 3D analysis, whereas disc
>   forging only requires a 2D analysis,

b) dimensional tolerances are far more important for blades (which are forged very close to their final shape), and

c) microstructural considerations are less important for blades, because the amount of work involved in the forging process means that a fine-grained structure is the norm.

The above points are all clearly important from the process modelling point of view. In particular, it may become necessary to include elastic deformation in the analysis to predict the final forging dimensions accurately (including spring-back when the dies are separated). The most important difference between compressor blade forging and disc forging, however, lies in the fact that the former is a 3D problem. Although local sections of a blade can be simulated as 2D planes, the practical problems associated with metal flow are three dimensional in nature. Clearly the complexities of the 3D analysis are greater than a 2D analysis, for instance in the 3D contact between the die and the workpiece.

A complication in the simulation of the forging of complex 3D parts is the need to create and manipulate 3D geometries. Ideally the dies should be designed on a CAD (Computer Aided Design) system and this geometry definition should serve as the reference for both machining the dies and for the simulation of the process. In practice, however, the interfaces between different systems for 3D geometric modelling are not straightforward and often errors can creep in due to different conventions for defining 3D surfaces.

Packages do exist which are capable of simulating the forging of complex 3D parts, including elastic deformation. In practice, however, the computing times currently required mean that simplications are needed to perform an analysis. In particular, the assumption of isothermal conditions will reduce the analysis time considerably.

3.2.3   OVERVIEW OF FORGING PROCESS
         MODELLING

In summary, the modelling of forging generally requires a large scale deformation simulation of the workpiece, combined with a thermal analysis and a small-scale deformation simulation for the die. For isothermal forging of discs, however, the thermal analysis is by definition not required and the analysis is 2D axisymmetric. This has resulted in forging modelling being used extensively for isothermally forged discs, including being a capability which has been used for publicity purposes.[11,12] An example of a 2D isothermal simulation is given in section 5.2. Forging modelling is also being used to simulate non-isothermal forging of discs with some success.

It should be stressed that the prediction of microstructure is critical to the successful use of forging modelling to reduce lead times and costs for jet engine discs. To do this it is essential for there to be the required level of understanding of how microstructures develop during forging, and how these microstructures vary with the parameters obtained from the analysis.

3D modelling of forging is, like the modelling of 3D casting, very intensive in the use of computing time. Its application has thus been limited at present, although some initial work has been done.[13,14] One alternative to numerical modelling is to use physical modelling, where a model material, such as a wax of similar characteristics to the hot metal, is deformed.[15] Although physical

181

modelling can be useful in showing the expected flow patterns in the forging, it is inherently limited in its scope. For instance one could not use physical modelling for predicting springback or microstructural development. Similarly, the scope for expert systems is seen as being limited by the constraints of the expertise on which the systems are based. Forging modelling will be widely adopted eventually for 3D parts because of the ability to simulate the forging of a new material, a new component shape or the use of a new processing route. For this to happen in practice, however, more powerful computers need to be more widely available, and also the logistics need to be sorted out of how to create and handle 3D geometries in computing systems. Extensive work is already in progress in both of these areas.

3.3 Heat Treatment

Heat treatment is used to ensure that forged discs have the microstructures and thus properties required to match the service conditions. Castings are also heat treated to achieve the required properties. Further, heat treatment is used to bring about stress relief in welded assemblies.

Nickel-base superalloy discs for jet engines are often oil quenched following solution heat treatment at temperatures of the order of 1050°C. This induces rapid cooling in the component, which is desirable to obtain the highest possible strength from the alloy. Due to the size of the discs involved, however, the rapid cooling of the surface of the disc relative to the centre induces stresses during quenching, which are sufficient to cause yielding and may be high enough for cracking to be a risk. The plastic strains induced result in there being residual stresses present in the heat treated disc which remain in the finished component, albeit at a lower level due to subsequent ageing and machining. These residual stresses can have an important influence on the fatigue behaviour of the component and thus its life in service. Residual stresses similarly occur in oil quenched titanium alloy discs.

Modelling heat treatment processes can thus reduce lead times for process development by predicting:

a)   the residual stresses,

b)   the formation of defects, such as cracks,

c)   the effect of the heat treatment process on microstructure, and

d)   distortion.

In general, problems do not occur in the heat treatment of cast turbine blades because the section thicknesses are low enough for there to be little variation in cooling rate in the component. What problems do occur tend to be more process related (eg uneven cooling due to the arrangement of the blades in a heat treated batch). The heat treatment of discs, however, merits particular attention due to the relatively large section thicknesses involved and the importance of the integrity of these components to engines in service.

To meet the needs of a model which represents the quenching of discs the package used requires the following characteristics:

a)   the ability to perform a thermal analysis, a small displacement elastic-plastic stress analysis and ideally a creep analysis,

b)   the handling of 2D axisymmetric geometries,

183

c)   boundary conditions in the thermal
      analysis which are highly non-linear
      functions of the difference in
      temperature between the disc and the
      quenching medium,

d)   boundary conditions which vary
      around the perimeter of the
      component,

e)   radiative and convective heat loss
      to the environment during transfer
      to the quench tank, and

f)   the ability to handle thermal and
      mechanical properties which are
      non-linear functions of temperature.

The creep analysis is ideally needed to
simulate stress relief in the ageing
operation. Also it can be used to predict
sagging due to the disc self weight while it
is undergoing solution heat treatment.[16]

Finite element based packages are available
which meet the above requirements and can be
used to predict cooling rates, stresses
developed during quenching, residual stresses
remaining after heat treatment, and stress
relief due to ageing.

Provided a stress based (or strain based)
failure criterion exists then cracking can be
predicted. Similarly microstructural effects
can be predicted if they can be related to
cooling rates (or some other derivative of
temperature). Furthermore, using judicious
adjustments to the assumed mechanical
properties, it is possible to predict the
gross effects of material removal (in terms of
residual stress relief and distortion).

Heat treatment modelling has been used to good
effect for discs to predict the features
listed as necessary for a model, namely

residual stresses, cracking,[17] microstructural effects and distortion. As for disc forging the fact that the analysis is 2D makes the computing task easier to handle. Computing times would typically run into a few hours on a state-of-the-art workstation.

4. DATA REQUIREMENTS FOR PROCESS MODELLING

It will be apparent from the previous discussion of the modelling needs for casting, forging and heat treatment that there is some commonality between the types of analysis required. In particular, the detailed modelling of all three processes requires a thermal analysis and a stress analysis. When consideration is given to how to model other manufacturing processes it becomes clear that there is more general commonality. This is perhaps not surprising when one considers how similar forming, extrusion and solid state welding are to forging, or how similar fusion welding, electric arc melting or electroslag refining are to casting.

The data required for fluid flow, thermal and stress analyses are covered in turn, with particular reference to the needs for modelling casting, forging and heat treatment, but the above points concerning the similarity of the needs for modelling other processes should be born in mind.

4.1 Fluid Flow Analysis

The immediate requirement is for data on viscosity for the fluid or fluids of interest. Clearly, in the case of casting modelling this will be strongly dependent on temperature, to the extent that at around the solidus temperature the material effectively becomes rigid.

If free surfaces are to represented, surface tension values are required. These will be dependent on temperature, and indeed for air casting may also depend on time due to reaction affects (i.e. oxidation).

Density values will be required for the metal, ideally for the liquid, mushy and solid states. Clearly these values will change rapidly during solidification.[18] Measurement of density may not be a simple matter, particularly for alloys which have a high melting point or are highly reactive.

For simulating mould filling there is likely to be a requirement for a thermal analysis coupled to the fluid flow simulation. The data needed for the thermal analysis are described below. It is worth noting, however, that the coupled thermal analysis will also need to be able to handle the moving boundary.

4.2 Thermal Analysis

In order to model heat conduction in the general case data are required for conductivity, density and specific heat, all of which can be combined to calculate diffusivity. These parameters generally vary with temperature, and clearly measurements taken to obtain data need to be made over the temperature range of relevance to the process being modelled. For alloys, it is generally assumed that the values for these data are primarily dependent on alloy chemistry and as such are not sensitive to the particular microstructure. For shell systems in investment castings, however, the values for the 'thermophysical' properties will depend on the detailed conditions used to 'invest' the shell and to fire it. For example, the particular particle size of the ceramic used for the 'stucco' may affect the packing density of the shell and thus its properties.

For casting modelling, data are needed for the

latent heat released on solidification. This is usually assumed to be released uniformly over the temperature range between the liquidus and solidus. In practice, however, for complex alloys this is a considerable simplication, and more sophisticated assumptions may be required if detailed predictions of microstructure are being performed. Heats of formation can also occur in the solid state, eg due to precipitation or changes in lattice structure, which may need to be allowed for in the thermal analysis depending on the amount of energy released and the accuracy required.

In addition to heat flow by conductivity, heat transfer by radiation may need to be catered for. For radiation calculations the relative emissivity of the surface of the object needs to be known. Measuring this is not straightforward at the high temperatures usually relevant to manufacturing processes.

Furthermore the characteristics of the surface may change during the process; for instance the shell of an investment casting often becomes discoloured once the metal is poured and large differences have been measured between the relative emissivity of the initial cream coloured shell and that for the 'dirty' shell at the end of the casting process.

Complex conditions exist at interfaces in casting, forging and heat treatment processes. In some cases a convective heat transfer coefficient may be assumed:

$$q = h A (T-T_o)$$

where:

q = heat flux

h = heat transfer coefficient

A = cross sectional area

T =   temperature of object of interest

T_o = temperature of cooling or heating
medium

This type of assumption is likely to be made
for heat transfer from a hot piece of metal
to a die, or for heat transfer from a hot
piece of metal to a quenching medium.   In
practice the difficulty lies in what value to
assume for h. Generally experiments involving
temperature measurements will be required,
where conditions are as similar as possible to
those present in the manufacturing process.
The art is to make appropriate approximations
to the situation being modelled. Thus for die
to workpiece heat transfer in forging,
different values for h might be assumed when
the die rests on the workpiece, when pressure
is applied to the die and lastly when the
forging operation is in progress. Similarly in
quenching components, experiments have shown
that the rate of heat transfer varies around
the perimeter of the component, and that this
variation depends of the design of the
particular quench facility used.

Another way of treating interfaces is to
introduce a thermal resistance at the
interface. This might be assumed for instance
for the mould/metal interface in a casting
model. One way which has been proposed to
represent this resistance in a finite element
model is to use a collapsed brick element of
zero thickness.[16]

4.3   Stress/deformation analyses

The stresses formed during casting and heat
treatment are caused by differential
contraction. This can either be because the
materials used have different expansion
coefficients (such as the mould and the metal
in a casting) or due to temperature gradients

present in the component (as for heat treatment).

The first property data required for a stress analysis are thus values for linear expansion coefficient over the range of temperatures of interest. This is complicated by phase changes. Thus to simulate shrinkage in castings the volume change at solidification must be quantified. This is difficult to measure, although for instance dilatometry work has been done on cast irons.[18] Solid state phase transformations may also be important. Certainly, the transformation to martensite in steels can have a marked effect on the stresses induced during heat treatment.[19]

In general values for linear expansion coefficient will be temperature dependent. It is worth noting that to calculate thermal stesses the expansion coefficient should be appropriate to the change in temperature experienced. This may require the calculation of instantaneous values of expansion coefficient rather than the values normally quoted which are referenced to ambient temperature. The relationship between the two is given below:

$$\alpha_{inst} = \frac{\alpha_2(T_2-T_o) - \alpha_1(T_1-T_o)}{(T_2-T_1)}$$

where:

α_{inst} = instantaneous value of linear expansion coefficient at temperature $(T_1+T_2)/2$

T_1,T_2 = temperatures

T_o = ambient temperature

α_1,α_2 = linear expansion coefficients referenced to ambient temperature for temperatures of T_1 and T_2, respectively

189

Where elastic displacements are included in an analysis, values for Young's modulus will be required over the temperature range of interest. High temperature measurements of Young's modulus are difficult to obtain, although the importance of elastic displacements is likely to be less the higher the temperature (because of the increased dominance of plasticity, or even creep, effects).

Plasticity needs to be catered for in deformation calculations for heat treatment (because plastic strains cause the residual stresses of interest) and forging, and also ideally for casting (to predict zones of recrystallisation, for instance). In heat treatment analyses, it is normally assumed that plasticity can be represented by a yield stress and work hardening. The yield stress will depend strongly on temperature and so will the work hardening rate. The stress analysis packages available allow the representation of complex work hardening relationships (often as a piecewise linear relationship between stress and strain), although simplications can clearly be made (the most extreme being to assume a zero work hardening rate).

For forging modelling the assumption is usually made that elastic displacements (at least in the forging itself) are negligible and that the deformation occurs due to plastic strains. Further the high temperatures involved mean that the 'flow' stress of the material is more strongly dependent on strain rate than strain. A common form of constitutive equation is based on a power law relationship between flow stress and strain rate, ie

$$\sigma = A\dot{\varepsilon}^n$$

where:

    S = flow stress

    $\dot{\varepsilon}$ = strain rate

    A,n = constants

This relationship is usually further refined so that in the general case, the flow stress is also dependent on strain (to represent work hardening and/or softening), temperature and microstructure or a 'structure related variable'. Thus:

$$\sigma = f(\varepsilon, \dot{\varepsilon}, T, S)$$

where:

    σ = flow stress

    ε = strain

    $\dot{\varepsilon}$ = strain rate

    T = temperature

    S = structure related variable

In practice a continuous relationship can be developed to provide the function in the above equation. The meaning of the structure related variable, however, is difficult to correlate directly with a microstructural characteristic. Rather it is necessary to carry out tests on material which is representative of the starting point in the forging operation. These tests are normally carried out for a matrix of conditions, with a constant strain rate and temperature for each test. The eventual microstructure present in the disc can then be related to the strains, strain rates and temperatures experienced in the forging operation.[20]

As well as affecting the mechanical properties

191

relevant to the forging operation, the
microstructure of the alloy will also
generally affect the mechanical properties
relevant to heat treatment modelling. This
means that in the most general case, the
mechanical properties relevant to quenching
will depend on the structure developed by
forging and the heat treatment preceding
quenching. In practice, however, analyses
based on mechanical property data for fully
heat treated material appear to provide a
reasonable prediction of residual stresses
remaining in a disc after oil quenching.

A final requirement for stress analyses may be
creep data,for instance to simulate the stress
relief during ageing, or the sagging during
solution heat treatment. Such data can be
generated by standard creep test techniques,
and clearly need to be carried at
representative temperatures on material in the
appropriate condition.

5.   EXAMPLES OF PROCESS MODELLING

5.1 Casting Modelling

5.1.1.   SCOPE OF EXERCISES

To assess the capabilities of packages to
simulate casting an exercise was carried out
on a jet engine casing investment casting.
This particular component was chosen for study
because the casting had been produced using
various different feeding arrangements, all of
which gave rise to different patterns in the
shrinkage present in the cast component. The
purpose of the exercise was:

> a)   to evaluate the ability of a
> straightforward thermal analysis to
> predict the locations of shrinkage,

b)   to demonstrate how the results might
      be displayed to best effect, and

c)   to gauge the effort involved in the
      modelling exercise.

5.1.2   DETAILS OF THE ANALYSIS

The geometry of the casting, running system
and feeding ring was created in a 'pre and
post-processing' package designed to interface
with finite element programs. The solid model
created is shown in fig 5. The invested shell
was added to this model to create a new solid
model, which thus incorporated the casting
itself, the running system, the feeding ring
and the shell.

From the solid model geometry including the
mould a finite element mesh was created. Due
to symmetry it was only necessary to perform
the analysis based on a segment comprising one
sixteenth of the geometry. The mesh created
for this segment is shown in fig 6 which
consists of 8 node quadrilateral elements.

It is worth noting that the geometry
definition and mesh creation took
approximately 10 man days to complete.

The analysis was performed using a general
purpose non-linear finite element package. An
input 'deck' was created for the analysis
using the data for the finite element mesh
shown in fig 6. Properties approximating to
those for a nickel-base superalloy were
assigned to those elements representing the
metal, and the elements representing the mould
had properties defined based on data for an
appropriate invested shell system. A latent
heat value was defined for the alloy and this
was assumed to be evolved uniformly over the
liquidus to solidus temperature range.

To perform the analysis it was assumed that
the mould was instanteously filled with metal

at a uniform temperature of 1477°C and that
the mould was also at a uniform temperature of
1050°C.

Heat loss by radiation to the environment was
assumed. The environment was taken to be at a
uniform and constant ambient temperature
(20°C). Radiative heat transfer between
different parts of the mould was not allowed
for. A further important simplification was
to assume a perfect thermal contact between
the mould and the metal (ie it was assumed
that there was no resistance to heat transfer
across the interface).

A transient thermal analysis was run until all
the regions in the metal had dropped below the
specified solidus temperature for the alloy.
The computing time required was equivalent to
18 hours on an IBM 3081 mainframe.

5.1.3   RESULTS FROM THE ANALYSIS

At each increment in the analysis (14 in all)
temperatures were computed throughout the
geometry. Examples of these plots are shown
in figs 7 and 8 for different times in the
analysis.

5.1.4   DISCUSSION

An objective of the exercise was to assess the
ability of a straight-forward thermal analysis
to predict the locations of shrinkage in the
casting. The simplest way of judging the
location of shrinkage from the results of a
thermal analysis is to search for regions
where metal at temperatures above the solidus
is surrounded by metal at temperatures below
the solidus. Applying this approach to the
results of this exercise would suggest that
the only enclosed shrinkage defect expected
would .be in the centre of the feeding ring

beneath the casing, as shown in figure 8. Although in practice shrinkage was found in the centre of the feeding ring it was also present at other positions within the casting. There are two possible reasons for this apparent inability to predict the position of shrinkage defects:

a)   the analysis is not sufficiently accurate in its prediction of the temperature distributions in the casting (due for instance to ignoring mould/metal interface thermal resistance or radiation between different parts of the mould), and

b)   studying isotherms is not sufficient as a method for predicting the position of shrinkage (due for instance to not representing the movement of metal due to volumetric contraction and gravity feeding).

It is considered necessary for a complete quantitative model of shrinkage to include the heat loss during mould filling and the geometry changes resulting from volumetric contraction in addition to performing the 'in-situ' thermal analysis.

The second objective of the exercise was to demonstrate how results can be displayed to best effect. Colour contour plots were produced to show temperature contours on the casting geometry. Through a selective choice of the colours used, it is possible to show the solidification sequence in the casting. Further, with moderate developments to the software, it would be possible to show parameters which can be derived from the temperatures predicted (eg temperature gradient, G, or growth rate, R). In general terms very effective displays of the results can be produced, the only limitations being on the performance of the hardware used (for instance contour plots took about 10 minutes to display) and on the interpretation of the data to provide variables relevant to microstructure and defect prediction. It is worth noting, however, that hardware performance improvements have already overcome the former limitations.

The third objective of the exercise was to gauge the effort required for the analysis. It was significant that the geometry creation and mesh generation took approximately 10 man days. This illustrates the importance of ensuring that geometries are only defined once, and that if a component is designed on a CAD system its geometry should not have to be recreated for modelling purposes. Furthermore, the creation of the mesh needs to utilise developments in automatic mesh generation. At present, it is both time consuming and requires expertise in numerical analysis to design a mesh relevant to the level of accuracy required from the analysis.

The second important performance measure is the computing time required for the analysis. The 18 hours analysis time on an IBM 3081 computer precludes the repeated calculations required for iterative process design, particularly if, as suggested previously, a simulation of mould filling is an additional requirement.

5.2 Forging Modelling

5.2.1   SCOPE OF EXERCISE

To demonstrate the current capabilities of forging modelling packages an exercise was carried out to simulate the forging of a shape with features similar to a turbine disc (fig 9).

The particular example chosen was designed to

test the ability of a forging model to represent extensive deformation of the workpiece together with complex conditions in the contact between the die and the workpiece (ie die contact and sliding with friction).

This exercise was chosen to demonstrate the ability to predict microstructural development, in particular the extent of recrystallisation in a nickel-base superalloy used for turbine discs in a jet engine.

5.2.1   DETAILS OF THE ANALYSIS

The geometry of the dies shown in fig 9 was defined in the forging modelling package used, which has its own routines for defining the dies and the workpiece. The geometry of the initial billet was also defined and a finite element mesh was created using 6 node triangular axisymmetric elements.

The analysis assumed the dies are rigid. Slip between the dies and the workpiece was allowed and a "friction-factor" of 0.2 was assumed (this being the proportion of the shear yield stress which is induced as a frictional shear stress).

The analysis assumed that the dies and workpiece remained at the same temperature throughout the analysis. This represents the conditions present in a modern isothermal forge.

Visco-plastic behaviour of the material being forged was assumed, with a complex constitutive equation being used to describe the relationship between flow stress, strain and strain rate. The extent of recrystallisation in the alloy was related to the strain and strain rate histories at a given point in the mesh.

197

Due to the extensive deformation occuring in the analysis it was necessary to create 15 new meshes. This is because the elements become so distorted that excessive numerical errors occur. (The particular program used for this exercise adapts the mesh when specified errors in the predicted results are exceeded).

The complete analysis required 3.0 hours on a Sun 3/60 workstation. It should be noted, however, that recent benchmarks indicate that the same analysis would take about 40 minutes on one of the current state-of-the-art workstations.

5.2.3   RESULTS FROM THE ANALYSIS

The predicted sequence of deformation during the forging operation is showing in figure 10. Contour plots can be obtained of strain, strain rate and microstructure at any stage of the forging operation. The predicted strain contours at the end of the simulation are shown in fig 11, and fig 12 shows the predicted levels of recrystallisation.

5.2.4   DISCUSSION

The results shown in figs 10, 11 and 12 illustrate the capabilities of current packages for modelling the isothermal forging of discs. Although the particular forging simulated was not made, similar work has been done for isothermally forged discs. This work has indicated that the extent of recrystallisation can be predicted to within the errors of the measurements of recrystallisation.

As well as being capable of predicting microstructures in isothermally forged discs, current packages can be used to simulate non-isothermal forging. These analyses are more complex and, apart from requiring longer

198

computer run times, the algorithms required
for prediction of microstructure are clearly
more complicated (involving the additional
variable of temperature).

6. <u>LEARNING POINTS FROM EXAMPLES, DISCUSSION
OF CURRENT LIMITATIONS AND FUTURE
DEVELOPMENTS</u>

The most important point to make concerning
the previous examples is that for many
two-dimensional (2D) problems, the potential
of process modelling can be realised now.
Thus, it is possible to simulate a proposed
isothermal forging route in about 1 to 3 hours
on a state-of-the-art workstation (costing
about £20000) and to predict:

a)   the microstructures which would be
     obtained,

b)   if defects are likely to be formed
     (such as laps), and

c)   if the equipment is capable of
     performing the task (eg. Is the
     predicted load within the capacity
     of the press?)

It should be noted, however, that to be able
to predict the above, it is not only necessary
to have suitable programs and computer, but it
is also necessary to have relevant material
property data, relevant characterisation of
the boundary conditions (eg. friction factor)
and finally relevant quantitative
understanding of the relationship between the
direct output from the program (ie. stress,
strain, strain rate histories) and the
microstructure or the conditions giving rise
to defects.

In practice, there is not only a dearth of
relevant material property and boundary
condition data for forging modelling (at least

199

in the open literature), but more particularly, there are limits to the data and understanding which are needed to quantify the evolution of microstructure and to provide criteria for the development of defects.

Although process modelling can be used in practice now, as described above, there are limitations which need to be overcome before process modelling is applied routinely in practice. These limitations, which are all illustrated to some extent in the previous examples, are discussed in turn below.

6.1 Access to and/or Creation of Geometry Data

An added reason why 2D analyses are simpler than 3D analyses is that the definition of the geometry is so much simpler. In particular, to define from a drawing the 2D shapes required for forging modelling (the billet and dies in particular) would take perhaps two hours on a CAD system, or using a pre- and postprocessing package for finite element analysis. The same process for a reasonably complex 3D casting, or indeed forging, would take approximately two days. This illustrates the importance of using CAD systems to design a 3D component in the first place, and ensuring that this component definition is accessible to the package or pre- and post-processor used for process modelling. Perhaps surprisingly this is rarely practical at present because of a combination of the following:

a) 3D component designs are still commonly only available as drawings, and

b) the interfaces between different packages which handle 3D geometries are not well developed. Indeed many problems have been experienced (in

terms of lost definition, corruption
of data, etc) in using IGES (Initial
Graphical Exchange Standard) for
transferring 3D geometries.

A further point to note is that the geometry
required for process modelling is rarely
simply the component geometry. For example:

a)   for forging modelling, the die
     shapes would need to be defined
     rather than the component shape,

b)   for casting modelling, the mould,
     running system, feeders and chills
     would need to be added to the
     component geometry. Even then,
     modifications to the component
     geometry are often made to
     compensate for shrinkage and/or
     distortion in the casting process,
     and

c)   it is rare for a component to be
     heat treated in its final component
     shape. Some machining is usually
     carried out after heat treatment.

6.2   Mesh Generation

The finite element and finite difference
numerical techniques used for process
modelling are approximate and the accuracy of
the results depends critically on the design
of the mesh used to discretise the geometry of
interest. A poorly designed mesh can not only
lead to numerical errors, but also may make
the analysis fail altogether. On the other
hand, if the mesh is too refined the computing
times may be excessive for the accuracy
required.

The current packages available largely require
the operator to use his or her judgement in

201

designing the mesh. This means that a fair degree of expertise is required, so that a mesh is developed which is appropriate to the type of problem being solved (ie. thermal, fluid flow or stress) and the particular geometry of the component.

The increased complexity of 3D shapes means that mesh design tends to be considerably more time consuming than for 2D shapes. Thus, about seven days were required to define the mesh for the casting modelling example described in the previous section. This compares with approximately two hours to define the die and mesh for the 2D forging modelling example. In both cases, expertise was required to ensure that the mesh design was appropriate to the analysis being performed.

The time and expertise required for mesh generation are likely to remain high in the near future. Significant developments in automatic and adaptive meshing, however, will in due course reduce the time and expertise required. These developments are discussed further in Section 6.4.

6.3 Material property data, boundary condition data, data for validation of models

The lack of relevant material property and boundary condition data has already been referred to in the context of forging modelling. This dearth of information is also apparent for casting modelling and heat treatment modelling, and indeed for all other manufacturing processes. Clearly this is, in part, a reflection of process modelling being in its relative infancy. The properties required, however, are also often difficult to measure and therefore expensive to obtain due to factors such as the high temperatures involved and the presence of multiple phases (e.g. liquid and solid in castings). Furthermore, the detailed composition of materials used in manufacturing processes, together with the design of the process itself are often considered as proprietary. These factors mean that where data are available they are often not widely publicised.

It is important to validate process models using tests of practical relevance. Without these tests, it will never be possible to show that the approximate methods used for analysis, together with the material property data and assumed boundary conditions, produce results of the required level of accuracy; or indeed that the analysis is representative of the problem in the first place. Careful consideration is needed when designing tests for validation purposes. In particular, it needs to be clear what is being validated. Is it the ability to predict microstructure in a forging or just the ability to predict the history of the metal flow? For castings, is it the ability to predict the growth of the solidification front, or to predict where shrinkage will occur?

There is certainly a shortage of data which can be used to validate process models. Obtaining such data is not only confused by the issue of what to validate a model against, but is also complicated by the difficulties and costs involved in obtaining the data, and by the inaccuracies or variability implicit in all measurement techniques.

6.4   Numerical Analysis Techniques

Currently available numerical analysis techniques, and indeed commercially available packages, are capable of doing the analyses required for forging, casting and heat treatment modelling. In practice, however, these packages are both inefficient in their use of computing time and are often based on incompatible numerical techniques.

The packages currently available are often general purpose packages which can be relatively inefficient, ie. slow, when applied to a particular problem. The use of packages designed solely to solve the problem of interest is one means of obtaining improvements in performance. Additional developments which look particularly promising include:

a) the use of conjugate gradient solvers,[16,21] and

b) the utilisation of parallel processing.

The latter development appears particularly promising, with the latest "top-of-the-range" workstations being produced with multiple processors. The limitation which needs to be overcome, however, is that most packages are not optimised to gain the benefits of parallel processing. These benefits can potentially be substantial, for instance a 7 Fold speed-up was demonstrated recently on a fluid flow problem using a board with 9 transputers, each of which costs approximately £800.[22]

A comprehensive model of a process often requires more than one type of analysis. Thus, in the general case, forging modelling requires a deformation and thermal analysis, and casting requires a fluid flow, thermal and deformation analysis. Commercial fluid flow packages, however, are largely based on finite difference (or "control volume") techniques, whereas commercial stress analysis, or deformation, packages are mainly based on finite element techniques. If the results of one type of analysis affect the assumptions of a later or coupled analysis this can cause complications. In particular, the types of element used for the two analyses will differ.

As a consequence the time consuming task of interpolating results from one analysis onto the mesh used for the other analysis will be necessary. It would be far simpler to be able to use the same type of analysis and thus mesh for both tasks. This may well be possible in the near future, either due to the development of finite difference codes which can handle finite element type meshes[23] or due to the development of fluid flow codes which use finite element techniques.[24] In practice, however, an interpolation routine may still be needed because the mesh suited to one type of analysis (eg. fluid flow) may not be appropriate to another type of analysis (eg. stress).

Two final areas of numerical methods development which have significant potential are automatic and adaptive meshing. Automatic meshing involves the automatic creation of a mesh for the numerical analysis which fills the space defined by the geometric model of the problem.[25] This is significant in that it relieves the operator of the task of mesh generation, but does not in itself mean that the mesh is appropriate to the problem being solved. Adaptive meshing, however, involves the modification of a previous mesh to attain a required level of accuracy from the analysis, based on error estimates made in an analysis performed using a previous mesh.[26]

The prize to be gained from these two developments is not increased computing speed (indeed adaptive meshing can result in the need for multiple analyses to obtain an appropriate mesh), but lies in the potential for a user who is unfamiliar with the detailed numerical analysis technique and meshing requirements to carry out an analysis to a prescribed accuracy. The implications to the widespread use of process modelling (and indeed the use of these numerical techniques in general) are considerable. So far, however, automatic and adaptive meshing

205

techniques have tended to be confined to 2D applications, including the non-linear problems of forging[27] and casting[28] modelling, although work has started on 3D problems.[29]

6.5 Computer Power

The computing times taken for the examples described in section 5 illustrate that process modelling is far from being a tool which can be used for rapid "what if" calculations. Indeed, a comprehensive simulation of the casting of a 3D component involving the mould filling simulation, as well as a thermal analysis of the casting once the mould is filled, could take many days on a current mainframe using current software. It is essential, however, to be aware of the general trends in computing power and the recent developments taking place. Over the long term large computers have shown increases in speed for algebraic problems of approximately two orders of magnitude per decade (fig 13). Thus, were these trends to continue, the 18 hour casting simulation described in Section 5.2 would take about 11 minutes in a decade's time. There are good reasons, however, to be more optimistic about improved computer performance. In particular,

a)   the increased availability of parallel processors, and

b)   the development of considerably faster, relatively low cost chips.[30]

In the workstation market, progress has been extremely rapid in the last five years. For instance, recent numerical benchmarks have shown that a state-of-the-art workstation two years ago was about ten times slower than the current state-of-the-art workstations (which cost roughly the same amount). For graphics operations, the development of graphics processors meant that the display of the 3D

solid model shown in fig.5 was thirty times
faster on one of the new workstations
available. The current competitive nature of
the workstation market, together with the
development of new chips, bodes well for the
continued development of even faster,
competitively priced, workstations.

6.6 "User-friendliness"

Important elements of the usefulness of
computer programs are how easy it is to learn
to use the program, as well as its reliability
or "robustness". Most numerical analysis
packages currently require a relatively long
period (perhaps three months) of use before a
user becomes familiar with the package.
Furthermore, the method for input of the data
and display of results is often cumbersome and
slow.

Some aspects of "user-friendliness" have
already been covered. In particular, the
following developments will significantly
improve matters:-

   a)   easy access to 3D geometries defined
        on CAD systems,

   b)   automatic mesh generation and
        adaptive meshing to achieve the
        required accuracy,

   c)   readily accessible material property
        and boundary condition data in the
        form required for the analyses,

   d)   rapid solvers with relevant
        interfaces between different types
        of analysis, and

   e)   faster computers.

In addition, however, thought needs to be
given on how to design the modelling system so

207

that data can be input and results displayed in the most relevant way to the user. This is clearly dependent on the user's skills and expectations. The software, however, needs to be developed so that process modelling is no longer the preserve of the numerical analyst (who is rarely familiar with or understands the manufacturing process and its problems), but becomes a tool that can be used by manufacturing process engineers to design the process based on their knowledge of the options to try. It is only when process modelling reaches this level and is used at the earliest possible stage of the design process that the real benefits in terms of reduced process development lead times and costs will be achieved.

7.   CONCLUDING REMARKS

This paper has provided an overview of how process modelling can fit into the design process and thus reduce the lead times for new product development. The requirements which need to be met to perform a successful analysis have been reviewed, specifically for forging, casting and heat treatment. Examples of forging and casting modelling have been given and the current limitations have been reviewed.

The important message is that the modelling of manufacturing processes in 2D is practical now, as is demonstrated by the regular use of process modelling to design isothermal forging routes for aero-engine discs. The limitations to the use of process modelling for 2D shapes are largely due to the limited material property and boundary condition data available, the limited quantitative understanding of microstructural development and defect formation and finally the validation of process modelling where "coupled" analyses (eg. deformation and thermal) are required. On a typical

208

state-of-the-art Unix workstation costing approximately £20000, a 2D isothermal forging simulation would take about 1-3 hours and a non-isothermal simulation might take approximately 2-6 hours, (although this could be greater depending on the numerical method used for the thermal analysis).

On the 3D front, here are various reasons why the application of process modelling is limited at present. Most prominent of these are the difficulties of generating 3D geometries and meshes and the computing times involved. There is no room for complacency, however, as significant developments are underway to address the current limitations. In particular, the developments of faster low cost workstations, parallel processing and automatic and adaptive meshing will all have a major impact on the practical viability of process modelling in 3D. The people who will realise the full potential of process modelling in 3D first will be those who, when the above developments have been made, have already addressed the issues of:

(a) how to interface with the CAD system used to design the component,

(b) understanding and quantifying the behaviour of the materials in the process, and

(c) quantifying the boundary conditions relevant to the process of interest.

An illustration of how process modelling is taking off is provided in Figures 14 and 15, which show the number of papers on casting and forging modelling, respectively, since 1969. A further indication of this is the number of books and conference proceedings published on this subject in the last few years (see the list of references in the appendix).

It is suggested that the trends shown in Figures 14 and 15 have many parallels with the early days of the use of linear elastic finite element stress analysis, and during the next few years it is expected that process modelling will become as intrinsic a part of the design process as stress analysis is now, particularly for high cost, critical components.

8. ACKNOWLEDGEMENTS

The author wishes to thank his colleagues, in particular G.J.S.Higginbotham, J.C.Hogg, M.A.King, P.J.Spence and P.D.Spilling, for their helpful comments on the text and assistance in preparing the paper.

9. REFERENCES

(1) J. NEWHOUSE: 'The sporty game', 1982, New York, A.A.Knopf.
(2) R. W. EVANS: in Proc. Sem. 'Numerical techniques', London, 6 Dec 1989, The Institute of Metals.
(3) D. FISHLOCK: 'Confining the failures to the computer', 13th June 1989, Financial Times.
(4) R. H. JEAL: Metals and Materials, 1988, 4, (9), 539-542.
(5) J. C. HOGG and H. WESTENGEN. 'Low pressure sand casting of magnesium alloys', to be presented at Inst. British Foundrymen (IBF) conf. on light alloy casting, Stratford-upon-Avon, 5-6 Dec 1989.
(6) G. J. S. HIGGINBOTHAM: Materials and Design, Part A, 1986, VII, (5), Part B, 1986, VII, (6), Part C, 1987, VIII, (1).
(7) W. KURZ and D.J. FISHER: 'Fundamentals of solidification', 1986, Aedermannsdorf, Switzerland, Trans-Tech Publications.
(8) M. McLEAN: 'Directionally solidified materials for high temperature service',

1983, London, The Metals Society.

(9) P. N. HANSEN and P. R. SAHM: in Proc. Conf. 'Modeling and control of casting and welding processes IV ', Palm Coast, FL, April 1988, (ed. A. F. Giamei and G. J. Abbaschian), 33-42, publ. Warrendale, PA, The Minerals, Metals and Materials Society.

(10) T. S. PIWONKA: JOM., 41, (2), 38-42.

(11) D. FISHLOCK: 'Massaging metal into shape', 5th Jan 1989, Financial Times.

(12) S. L. RAMSEY and M. A. RIPEPI: in Proc. Conf. 'Near net shape manufacturing'. Columbus, OH, Nov 1988, (ed. P. W. Lee and B. L. Ferguson), 75-77, 1988, Metals Park, OH, ASM International.

(13) I. PILLINGER et al: Proc. Inst. Mech. Engrs., 1985, 199, (C4), 319-324.

(14) G. SURDON and J. L. CHENOT: Int. J. Numer. Methods. Eng., 1987, 24, (11), 2107-2117.

(15) T. WANHEIM, V. MAEGAARD and J. DANCKERT: Advanced Technology of Plasticity, 1984, II, 984-996.

(16) R. O. STAFFORD: in Proc. Conf. 'Aerospace materials process modelling'. Cesme, Turkey, Oct 1987, Report AGARD-CP-426, (paper 13), AGARD, Neuilly sur Seine, France, 1988.

(17) R. A. WALLIS et al: JOM., 1989, 41, (2), 35-37.

(18) R. HUMMER: Cast Metals, 1988, 1, (2), 62-68.

(19) A. J. FLETCHER: 'Thermal stress and strain generation in heat treatment', 1989, Barking, Essex, Elsevier Science Publishers Ltd.

(20) R. W. EVANS: in Proc. Conf. 'Aerospace materials process modelling'. Cesme, Turkey, Oct 1987, Report AGARD-CP-426, (paper 4), AGARD, Neuilly sur Seine, France, 1988.

(21) M. PAPADRAKAKIS and C. J. GANTES: Int. J. Numer. Methods. Eng., 1989, 28, (6), 1299-1316.

(22) M. CROSS: Presentation to Inst. Metals symposium on "simulating solidification in casting processes", 29-30 Nov 1988, London.

(23) R. D. LONSDALE and R.WEBSTER: in Proc. Conf. 'Numerical methods in laminar and turbulent flow ', Swansea, July 1989, (ed. C. Taylor et al), publ. 1989, Vol.6, Part.2, Swansea, Pineridge Press.

(24) R. LOHNER: Int. J. Numer. Methods. Eng., 1987, 24, (9), 1741-1756.

(25) M. A. YERRY and M. S. SHEPHARD: Computers and Structures, 1985, 20, 173-180.

(26) I. BABUSKA and W. RHEINBOLDT: SIAM Jnl.Num.Analysis, 1978, 15, 736-754.

(27) O. C. ZIENKIEWICZ and G. C. HUANG: in Proc. Conf. 'Numiform 89', Fort Collins, Colo, June 1989, (ed. E. G. Thompson et al.), 3-10, 1989, Rotterdam, A.A.Balkema.

(28) R. W. LEWIS, H. C. HUANG and A. S. USMANI: in Proc. Eurotherm Seminar No.6 on heat transfer in phase change problems, Delft, The Netherlands, Oct 1988.

(29) J. PERAIRE et al: Int. J. Numer. Methods. Eng., 1988, 26, (10), 2135-2159.

(30) W. R. FROELICH and E. MAYER: Supercomputing, 1989, 2, (4), 4-7.

APPENDIX: FURTHER READING ON PROCESS MODELLING

Process modelling in general

Proc. sessions. 'Process modeling - fundamentals and application to metals', 1978 and 1979, publ. 1980, Metals Park, OH, American Society for Metals

Proc. seminar. 'Computer simulation in materials science', Lake Buena Vista, FL, Oct 1986, (ed. R. J. Arsenault et al), publ. 1986, Metals Park, OH, ASM International.

Proc. Conf. 'Aerospace materials process modelling', Cesme, Turkey, Oct 1987, publ. Report AGARD-CP-426, 1988, Neuilly sur Seine, France, AGARD.

Proc. Conf. ' Mathematical models for metals and materials applications', Sutton Coldfield, Oct 1987, publ. 1987, London, The Institute of Metals.

JOM., 41, (2), Feb 1989. - Contains series of papers on process modelling.

Proc. Conf. 'Numerical methods in thermal problems', Swansea, July 1989, (ed. R. W. Lewis and K. Morgan), publ. Vol.VI,Parts 1 and 2, Swansea, Pineridge Press. - See also proceedings for 4th and 5th conferences from same publisher in 1985 and 1987, respectively.

Heat treatment modelling

Proc. Conf. 'Calculation of internal stresses in the heat treatment of metallic materials ', Linkoping, Sweden, May 1984, (ed. E. Attebo and T. Ericsson), publ. 1984, Sweden, Linkoping University.

C. R. BOER et al: (1986) - see forging modelling section (also contains some work on heat treatment modelling)

A. J. FLETCHER: 'Thermal stress and strain generation in heat treatment', 1989, Barking, Essex, Elsevier Science Publishers Ltd.

Forging modelling

Proc. Conf. 'Numerical methods in industrial forming processes', Swansea, July 1982, (ed. J. F. T. Pittman et al), publ. 1982, Swansea, Pineridge Press.

Proc. Lecture series 'Process modelling applied to metal forming and thermomechanical processing', Oslo, Norway and Paris, France, Oct 1984, publ. Report AGARD-LS-137, 1984, Neuilly sur Seine, France, AGARD.

C.R. BOER et al: 'Process modelling of metal forming and thermomechanical treatment',1986, Berlin, Springer-Verlag.

Proc. Conf. 'Numiform 86', Gothenburg, Sweden, Aug 1986, publ. Int. J. Numer. Methods. Eng., 1988, 25, (1).

Proc. Euromech 233 colloquium. 'Modelling of metal forming processes', Sophia Antipolis, France, Aug 1988, (ed. J. L. Chenot and E. Onate), publ. 1988, Dordrecht, The Netherlands, Kluwer Academic Publishers.

Proc. Conf. 'Numiform 89', Fort Collins, Colo, June 1989, (ed. E. G. Thompson et al.), publ. 1989, Rotterdam, A.A.Balkema.

S. KOBAYASHI, S-I. OH and T. ALTAN: 'Metal forming and the finite element method', 1989, Oxford, Oxford University Press.

Casting modelling

Proc. Conf. 'Modeling of casting and welding processes II', Henniker, New Hampshire, July/Aug 1983, publ. 1984, Warrendale, PA, The Metallurgical Society of AIME.

P. R. SAHM and P. N. HANSEN: 'Numerical simulation and modelling of casting and solidification processes for foundry and cast-house', 1984, Comite International des Associations Techniques de Fonderie (CIATF).

Proc. 3rd Conf. 'Modeling and control of casting and welding processes', Santa Barbera, California, Jan 1986, publ. 1986, Warrendale, PA, The Metallurgical Society of AIME.

Proc. Symposium. 'State of the art of computer simulation of casting and solidification processes', Strasbourg, France, June 1986, (ed. H. Fredriksson), publ. 1986, Les Ulis, France, Editions de Physique.

Proc. Conf. 'Solidification Processing 1987', Sheffield, Sept 1987, publ. 1988, London, The Institute of Metals.

Proc. Conf. 'Modeling and control of casting and welding processes IV ', Palm Coast, FL, April 1988, (ed. A. F. Giamei and G. J. Abbaschian), publ. 1988, Warrendale, PA, The Minerals, Metals and Materials Society.

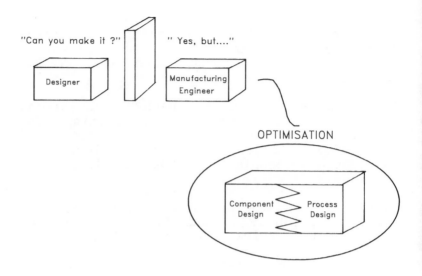

Fig.1.   Integration required between
           component and process design.

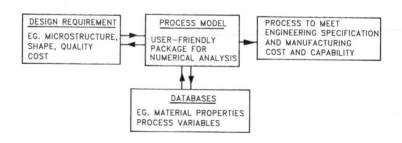

Fig.2.   Overview of process optimisation
           through process modelling.

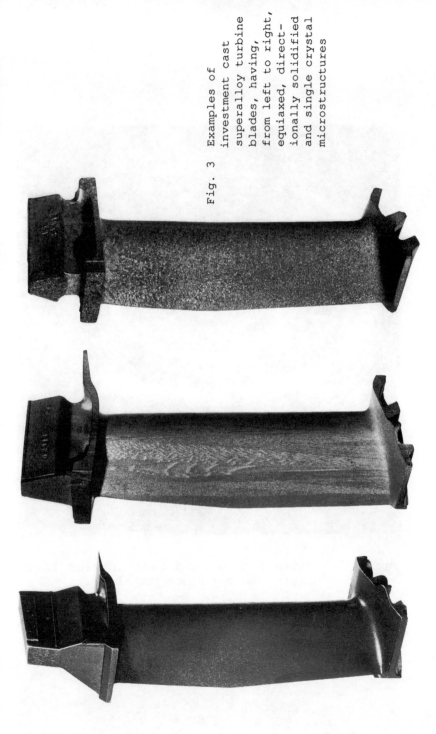

Fig. 3 Examples of investment cast superalloy turbine blades, having, from left to right, equiaxed, directionally solidified and single crystal microstructures

Fig.4.   Example of a sand cast intermediate casing in a magnesium alloy.

Fig.5.   Solid model of jet engine casing
          casting, including feeder ring and
          running system.

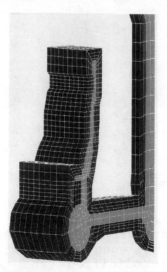

Fig.6.   Finite element mesh of 1/16th
          segment of jet engine casing
          casting.

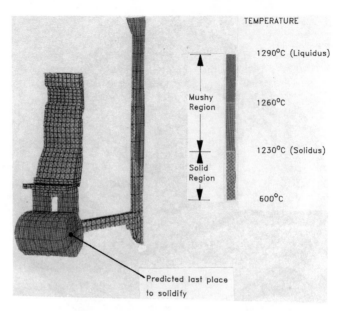

Fig.7.   Temperature contours predicted in
         casting when component itself is
         about 50% solid.

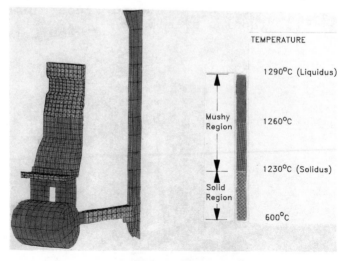

Fig.8.   Temperature contours predicted in
         casting when component itself is
         fully solidified (the metal in the
         feeding ring still being in the
         mushy state).

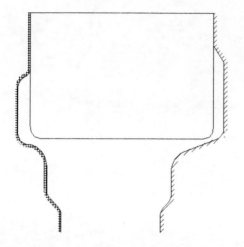

Fig.9. Forging modelling example.

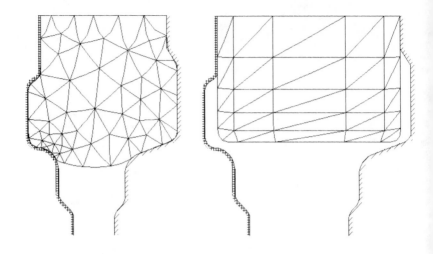

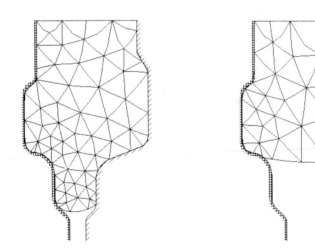

Fig.10.   Predicted sequence during forging
           operation.

222

Effective Strain

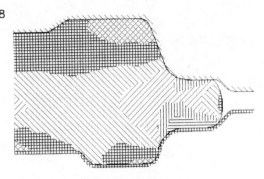

Less than 0.30
0.30 to 0.55
0.55 to 0.79
0.79 to 1.03
1.03 to 1.28
Greater than 1.28

Fig.11.   Contours of predicted strain for
           forging modelling example.

Fraction Recrystallised

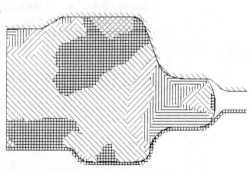

Less than 0.12
0.12 to 0.23
0.23 to 0.34
0.34 to 0.45
0.45 to 0.56
Greater than 0.56

Fig.12.   Contours of predicted levels of
           recrystallisation   for   forging
           modelling example.

223

EFFECTIVE SPEED (MFLOPS)

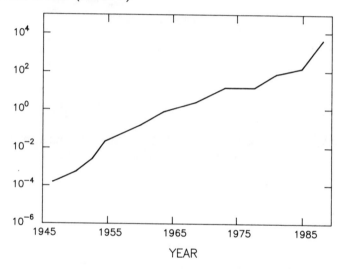

Fig.13.   Increases in computing speed since
           mid 1940's (data for "scientific
           computers"   taken   from   Financial
           Times, 10th April 1986).

NUMBER OF REFERENCES

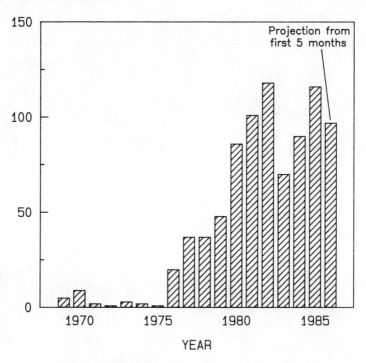

Fig.14.   Number of references to "numerical
           simulation of casting" in Metals
           Abstracts since 1969.

NUMBER OF REFERENCES

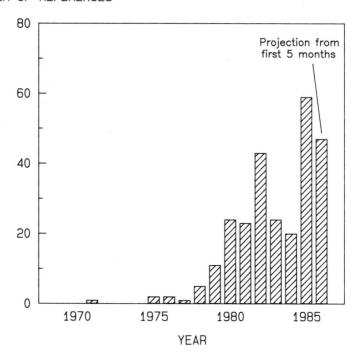

Fig.15. Number of references to "numerical simulation of forging" in Metals Abstracts since 1969.